INVESTIGATIONS IN NUMBER, DATA, AND SPACE®

Multiplication and Division

Arrays and Shares

Grade 4

Also appropriate for Grade 5

Karen Economopoulos
Cornelia Tierney
Susan Jo Russell

Developed at TERC, Cambridge, Massachusetts

Dale Seymour Publications®

White Plains, New York

The *Investigations* curriculum was developed at TERC (formerly
Technical Education Research Centers) in collaboration with Kent State
University and the State University of New York at Buffalo. The work was
supported in part by National Science Foundation Grant No. ESI-9050210.
TERC is a nonprofit company working to improve mathematics and science
education. TERC is located at 2067 Massachusetts Avenue, Cambridge,
MA 02140.

**This project was supported, in part,
by the**
National Science Foundation
Opinions expressed are those of the authors
and not necessarily those of the Foundation

Managing Editor: Catherine Anderson

Series Editor: Beverly Cory

Revision Team: Laura Marshall Alavosus, Ellen Harding, Patty Green Holubar,
Suzanne Knott, Beverly Hersh Lozoff

ESL Consultant: Nancy Sokol Green

Production/Manufacturing Director: Janet Yearian

Production/Manufacturing Coordinator: Joe Conte

Design Manager: Jeff Kelly

Design: Don Taka

Illustrations: Margaret Sanfilippo, Hollis Burkhart

Cover: Bay Graphics

Composition: Archetype Book Composition

This book is published by Dale Seymour Publications®, an imprint of
Addison Wesley Longman, Inc.

Dale Seymour Publications
10 Bank Street
White Plains, NY 10602
Customer Service: 1-800-872-1100

**DALE
SEYMOUR
PUBLICATIONS®**

Order number DS43891
ISBN 1-57232-744-8
2 3 4 5 6 7 8 9 10-ML-02 01 00 99 98

 Printed on Recycled Paper

TERC

Principal Investigator   Susan Jo Russell

Co-Principal Investigator   Cornelia Tierney

Director of Research and Evaluation   Jan Mokros

Curriculum Development
Joan Akers
Michael T. Battista
Mary Berle-Carman
Douglas H. Clements
Karen Economopoulos
Ricardo Nemirovsky
Andee Rubin
Susan Jo Russell
Cornelia Tierney
Amy Shulman Weinberg

Evaluation and Assessment
Mary Berle-Carman
Abouali Farmanfarmaian
Jan Mokros
Mark Ogonowski
Amy Shulman Weinberg
Tracey Wright
Lisa Yaffee

Teacher Support
Rebecca B. Corwin
Karen Economopoulos
Tracey Wright
Lisa Yaffee

Technology Development
Michael T. Battista
Douglas H. Clements
Julie Sarama
Andee Rubin

Video Production
David A. Smith

Administration and Production
Amy Catlin
Amy Taber

**Cooperating Classrooms
for This Unit**
Kay O'Connell
Linda Verity
*Arlington Public Schools
Arlington, MA*

Angela Philactos
Michele DeSilva
Jim Samson
*Boston Public Schools
Boston, MA*

Consultants and Advisors
Elizabeth Badger
Deborah Lowenberg Ball
Marilyn Burns
Ann Grady
Joanne M. Gurry
James J. Kaput
Steven Leinwand
Mary M. Lindquist
David S. Moore
John Olive
Leslie P. Steffe
Peter Sullivan
Grayson Wheatley
Virginia Woolley
Anne Zarinnia

Graduate Assistants
Joanne Caniglia
Pam DeLong
Carol King
Kent State University

Rosa Gonzalez
Sue McMillen
Julie Sarama
Sudha Swaminathan
State University of New York at Buffalo

Revisions and Home Materials
Cathy Miles Grant
Marlene Kliman
Margaret McGaffigan
Megan Murray
Kim O'Neil
Andee Rubin
Susan Jo Russell
Lisa Seyferth
Myriam Steinback
Judy Storeygard
Anna Suarez
Cornelia Tierney
Carol Walker
Tracey Wright

CONTENTS

TEACHER NOTES

WHERE TO START

The first-time user of *Arrays and Shares* should read the following:

When you next teach this same unit, you can begin to read more of the background. Each time you present the unit, you will learn more about how your students understand the mathematical ideas.

Investigations in Number, Data, and Space® is a K–5 mathematics curriculum with four major goals:

- to offer students meaningful mathematical problems
- to emphasize depth in mathematical thinking rather than superficial exposure to a series of fragmented topics
- to communicate mathematics content and pedagogy to teachers
- to substantially expand the pool of mathematically literate students

The *Investigations* curriculum embodies a new approach based on years of research about how children learn mathematics. Each grade level consists of a set of separate units, each offering 2–8 weeks of work. These units of study are presented through investigations that involve students in the exploration of major mathematical ideas.

Approaching the mathematics content through investigations helps students develop flexibility and confidence in approaching problems, fluency in using mathematical skills and tools to solve problems, and proficiency in evaluating their solutions. Students also build a repertoire of ways to communicate about their mathematical thinking, while their enjoyment and appreciation of mathematics grows.

The investigations are carefully designed to invite all students into mathematics—girls and boys, members of diverse cultural, ethnic, and language groups, and students with different strengths and interests. Problem contexts often call on students to share experiences from their family, culture, or community. The curriculum eliminates barriers—such as work in isolation from peers, or emphasis on speed and memorization—that exclude some students from participating successfully in mathematics. The following aspects of the curriculum ensure that all students are included in significant mathematics learning:

- Students spend time exploring problems in depth.
- They find more than one solution to many of the problems they work on.

- They invent their own strategies and approaches, rather than rely on memorized procedures.
- They choose from a variety of concrete materials and appropriate technology, including calculators, as a natural part of their everyday mathematical work.
- They express their mathematical thinking through drawing, writing, and talking.
- They work in a variety of groupings—as a whole class, individually, in pairs, and in small groups.
- They move around the classroom as they explore the mathematics in their environment and talk with their peers.

While reading and other language activities are typically given a great deal of time and emphasis in elementary classrooms, mathematics often does not get the time it needs. If students are to experience mathematics in depth, they must have enough time to become engaged in real mathematical problems. We believe that a minimum of 5 hours of mathematics classroom time a week—about an hour a day—is critical at the elementary level. The scope and pacing of the *Investigations* curriculum are based on that belief.

We explain more about the pedagogy and principles that underlie these investigations in Teacher Notes throughout the units. For correlations of the curriculum to the NCTM Standards and further help in using this research-based program for teaching mathematics, see the following books, available from Dale Seymour Publications:

- *Implementing the* Investigations in Number, Data, and Space® *Curriculum*
- *Beyond Arithmetic: Changing Mathematics in the Elementary Classroom* by Jan Mokros, Susan Jo Russell, and Karen Economopoulos

This book is one of the curriculum units for *Investigations in Number, Data, and Space.* In addition to providing part of a complete mathematics curriculum for your students, this unit offers information to support your own professional development. You, the teacher, are the person who will make this curriculum come alive in the classroom; the book for each unit is your main support system.

Although the curriculum does not include student textbooks, reproducible sheets for student work are provided in the unit and are also available as Student Activity Booklets. Students work actively with objects and experiences in their own environment and with a variety of manipulative materials and technology, rather than with a book of instruction and problems. We strongly recommend use of the overhead projector as a way to present problems, to focus group discussion, and to help students share ideas and strategies.

Ultimately, every teacher will use these investigations in ways that make sense for his or her particular style, the particular group of students, and the constraints and supports of a particular school environment. Each unit offers information and guidance for a wide variety of situations, drawn from our collaborations with many teachers and students over many years. Our goal in this book is to help you, a professional educator, implement this curriculum in a way that will give all your students access to mathematical power.

Investigation Format

The opening two pages of each investigation help you get ready for the work that follows.

What Happens This gives a synopsis of each session or block of sessions.

Mathematical Emphasis This lists the most important ideas and processes students will encounter in this investigation.

What to Plan Ahead of Time These lists alert you to materials to gather, sheets to duplicate, transparencies to make, and anything else you need to do before starting.

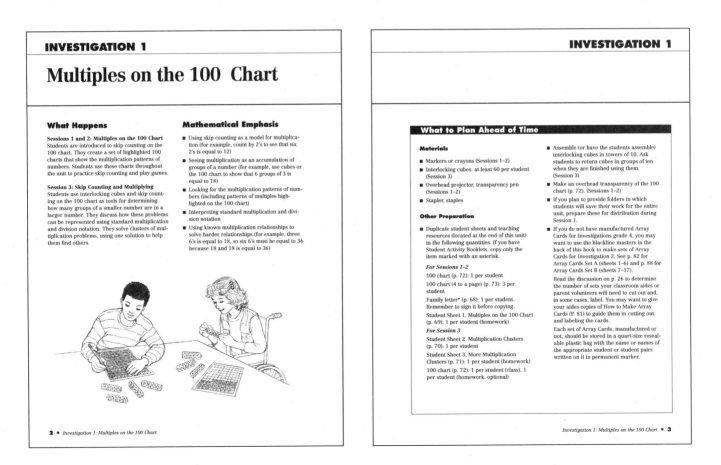

Sessions Within an investigation, the activities are organized by class session, a session being at least a one-hour math class. Sessions are numbered consecutively through an investigation. Often several sessions are grouped together, presenting a block of activities with a single major focus.

When you find a block of sessions presented together—for example, Sessions 1, 2, and 3—read through the entire block first to understand the overall flow and sequence of the activities. Make some preliminary decisions about how you will divide the activities into three sessions for your class, based on what you know about your students. You may need to modify your initial plans as you progress through the activities, and you may want to make notes in the margins of the pages as reminders for the next time you use the unit.

Be sure to read the Session Follow-Up section at the end of the session block to see what homework assignments and extensions are suggested as you make your initial plans.

While you may be used to a curriculum that tells you exactly what each class session should cover, we have found that the teacher is in a better position to make these decisions. Each unit is flexible and may be handled somewhat differently by every teacher. Although we provide guidance for how many sessions a particular group of activities is likely to need, we want you to be active in determining an appropriate pace and the best transition points for your class. It is not unusual for a teacher to spend more or less time than is proposed for the activities.

Ten-Minute Math At the beginning of some sessions, you will find Ten-Minute Math activities. These are designed to be used in tandem with the investigations, but not during the math hour. Rather, we hope you will do them whenever you have a spare 10 minutes—maybe before lunch or recess, or at the end of the day.

Ten-Minute Math offers practice in key concepts, but not always those being covered in the unit. For example, in a unit on using data, Ten-Minute Math must revisit geometric activities done earlier in the year. Complete directions for the suggested activities are included at the end of each unit.

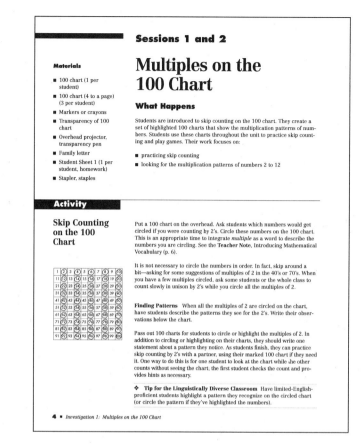

Activities The activities include pair and small-group work, individual tasks, and whole-class discussions. In any case, students are seated together, talking and sharing ideas during all work times. Students most often work cooperatively, although each student may record work individually.

Choice Time In most units, some sessions are structured with activity choices. In these cases, students may work simultaneously on different activities focused on the same mathematical ideas. Students choose which activities they want to do, and they cycle through them.

You will need to decide how to set up and introduce these activities and how to let students make their choices. Some teachers present them as station activities, in different parts of the room. Some list the choices on the board as reminders or have students keep their own lists.

Tips for the Linguistically Diverse Classroom At strategic points in each unit, you will find concrete suggestions for simple modifications of the teaching strategies to encourage the participation of all students. Many of these tips offer alternative ways to elicit critical thinking from students at varying levels of English proficiency, as well as from other students who find it difficult to verbalize their thinking.

The tips are supported by suggestions for specific vocabulary work to help ensure that all students can participate fully in the investigations. The Preview for the Linguistically Diverse Classroom lists important words that are assumed as part of the working vocabulary of the unit. Second-language learners will need to become familiar with these words in order to understand the problems and activities they will be doing. These terms can be incorporated into students' second-language work before or during the unit. Activities that can be used to present the words are found in the appendix, Vocabulary Support for Second-Language Learners. In addition, ideas for making connections to students' languages and cultures, included on the Preview page, help the class explore the unit's concepts from a multicultural perspective.

Session Follow-Up: Homework In *Investigations,* homework is an extension of classroom work. Sometimes it offers review and practice of work done in class, sometimes preparation for upcoming activities, and sometimes numerical practice that revisits work in earlier units. Homework plays a role both in supporting students' learning and in helping inform families about the ways in which students in this curriculum work with mathematical ideas.

Depending on your school's homework policies and your own judgment, you may want to assign more homework than is suggested in the units. For this purpose you might use the practice pages, included as blackline masters at the end of this unit, to give students additional work with numbers.

For some homework assignments, you will want to adapt the activity to meet the needs of a variety of students in your class: those with special needs, those ready for more challenge, and second-language learners. You might change the numbers in a problem, make the activity more or less complex, or go through a sample activity with those who need extra help. You can modify any

student sheet for either homework or class use. In particular, making numbers in a problem smaller or larger can make the same basic activity appropriate for a wider range of students.

Another issue to consider is how to handle the homework that students bring back to class—how to recognize the work they have done at home without spending too much time on it. Some teachers hold a short group discussion of different approaches to the assignment; others ask students to share and discuss their work with a neighbor; still others post the homework around the room and give students time to tour it briefly. If you want to keep track of homework students bring in, be sure it ends up in a designated place.

Session Follow-Up: Extensions Sometimes in Session Follow-Up, you will find suggested extension activities. These are opportunities for some or all students to explore a topic in greater depth or in a different context. They are not designed for "fast" students; mathematics is a multifaceted discipline, and different students will want to go further in different investigations. Look for and encourage the sparks of interest and enthusiasm you see in your students, and use the extensions to help them pursue these interests.

Excursions Some of the *Investigations* units include excursions—blocks of activities that could be omitted without harming the integrity of the unit. This is one way of dealing with the great depth and variety of elementary mathematics— much more than a class has time to explore in any one year. Excursions give you the flexibility to make different choices from year to year, doing the excursion in one unit this time, and next year trying another excursion.

Materials

A complete list of the materials needed for teaching this unit follows the unit overview. Some of these materials are available in kits for the *Investigations* curriculum. Individual items can also be purchased from school supply dealers.

Classroom Materials In an active mathematics classroom, certain basic materials should be available at all times: interlocking cubes, pencils, unlined paper, graph paper, calculators, things to count with, and measuring tools. Some activities in this curriculum require scissors and glue sticks or tape. Stick-on notes and large paper are also useful materials throughout.

So that students can independently get what they need at any time, they should know where these materials are kept, how they are stored, and how they are to be returned to the storage area. For example, interlocking cubes are best stored in towers of ten; then, whatever the activity, they should be returned to storage in groups of ten at the end of the hour. You'll find that establishing such routines at the beginning of the year is well worth the time and effort.

Student Sheets and Teaching Resources Student recording sheets and other teaching tools needed for both class and homework are provided as reproducible blackline masters at the end of each unit.

We think it's important that students find their own ways of organizing and recording their work. They need to learn how to explain their thinking with both drawings and written words, and how to organize their results so someone else can understand them. For this reason, we deliberately do not provide student sheets for every activity. Regard-

less of the form in which students do their work, we recommend that they keep their work in a mathematics folder, notebook, or journal so that it is always available to them for reference.

Student Activity Booklets These booklets contain all the sheets each student will need for individual work, freeing you from extensive copying (although you may need or want to copy the occasional teaching resource on transparency film or card stock, or make extra copies of a student sheet).

Calculators and Computers Calculators are used throughout Investigations. Many of the unity recommend that you have at least one calculator for each pair. You will find calculator activities, plus Teacher Notes discussing this important mathematical tool, in an early unit at each grade level. It is assumed that calculators will be readily available for student use.

Computer activities are offered at all grade levels. How you use the computer activities depends on the number of computers you have available.

Technology in the Curriculum discusses ways to incorporate the use of calculators and computers into classroom activities.

Children's Literature Each unit offers a list of related children's literature that can be used to support the mathematical ideas in the unit. Sometimes an activity is based on a specific children's book, with suggestions for substitutions where practical. While such activities can be adapted and taught without the book, the literature offers a rich introduction and should be used whenever possible.

Investigations at Home It is a good idea to make your policy on homework explicit to both students and their families when you begin teaching with *Investigations*. How frequently will you be assigning homework? When do you expect homework to be completed and brought back to school? What are your goals in assigning homework? How independent should families expect their children to be? What should the parent's or guardian's role be? The more explicit you can be about your expectations, the better the homework experience will be for everyone.

Investigations at Home (a booklet available separately for each unit, to send home with students) gives you a way to communicate with families about the work students are doing in class. This booklet includes a brief description of every session, a list of the mathematics content emphasized in each investigation, and a discussion of each homework assignment to help families more effectively support their children. Whether or not you are using the *Investigations* at Home booklets, we expect you to make your own choices about homework assignments. Feel free to omit any and to add extra ones you think are appropriate.

Family Letter A letter that you can send home to students' families is included with the blackline masters for each unit. Families need to be informed about the mathematics work in your classroom; they should be encouraged to participate in and support their children's work. A reminder to send home the letter for each unit appears in one of the early investigations. These letters are also available separately in Spanish, Vietnamese, Cantonese, Hmong, and Cambodian.

Help for You, the Teacher

Because we believe strongly that a new curriculum must help teachers think in new ways about mathematics and about their students' mathematical thinking processes, we have included a great deal of material to help you learn more about both.

About the Mathematics in This Unit This introductory section summarizes the critical information about the mathematics you will be teaching. It describes the unit's central mathematical ideas and the ways students will encounter them through the unit's activities.

About the Assessment in This Unit This introductory section highlights Teacher Checkpoints and assessment activities contained in the unit. It offers questions to stimulate your assessment as you observe the development of students' mathematical thinking and learning.

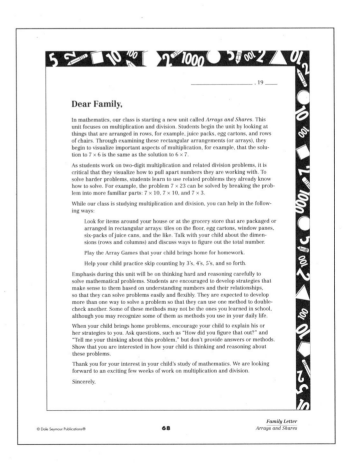

_____ , 19___

Dear Family,

In mathematics, our class is starting a new unit called *Arrays and Shares*. This unit focuses on multiplication and division. Students begin the unit by looking at things that are arranged in rows, for example, juice packs, egg cartons, and rows of chairs. Through examining these rectangular arrangements (or arrays), they begin to visualize important aspects of multiplication, for example, that the solution to 7 × 6 is the same as the solution to 6 × 7.

As students work on two-digit multiplication and related division problems, it is critical that they visualize how to pull apart numbers they are working with. To solve harder problems, students learn to use related problems they already know how to solve. For example, the problem 7 × 23 can be solved by breaking the problem into more familiar parts: 7 × 10, 7 × 10, and 7 × 3.

While our class is studying multiplication and division, you can help in the following ways:

  Look for items around your house or at the grocery store that are packaged or arranged in rectangular arrays: tiles on the floor, egg cartons, window panes, six-packs of juice cans, and the like. Talk with your child about the dimensions (rows and columns) and discuss ways to figure out the total number.

  Play the Array Games that your child brings home for homework.

  Help your child practice skip counting by 3's, 4's, 5's, and so forth.

Emphasis during this unit will be on thinking hard and reasoning carefully to solve mathematical problems. Students are encouraged to develop strategies that make sense to them based on numbers and their relationships, so that they can solve problems easily and flexibly. They are expected to develop more than one way to solve a problem so that they can use one method to double-check another. Some of these methods may not be the ones you learned in school, although you may recognize some of them as methods you use in your daily life.

When your child brings home problems, encourage your child to explain his or her strategies to you. Ask questions, such as "How did you figure that out?" and "Tell me your thinking about this problem," but don't provide answers or methods. Show that you are interested in how your child is thinking and reasoning about these problems.

Thank you for your interest in your child's study of mathematics. We are looking forward to an exciting few weeks of work on multiplication and division.

Sincerely,

Teacher Notes These reference notes provide practical information about the mathematics you are teaching and about our experience with how students learn. Many of the notes were written in response to actual questions from teachers or to discuss important things we saw happening in the field-test classrooms. Some teachers like to read them all before starting the unit, then review them as they come up in particular investigations.

Dialogue Boxes Sample dialogues demonstrate how students typically express their mathematical ideas, what issues and confusions arise in their thinking, and how some teachers have guided class discussions.

These dialogues are based on the extensive classroom testing of this curriculum; many are word-for-word transcriptions of recorded class discussions. They are not always easy reading; sometimes it may take some effort to unravel what the students are trying to say. But this is the value of these dialogues; they offer good clues to how your students may develop and express their approaches and strategies, helping you prepare for your own class discussions.

Where to Start You may not have time to read everything the first time you use this unit. As a first-time user, you will likely focus on understanding the activities and working them out with your students. Read completely through all the activities before starting to present them. Also read those sections listed in the Contents under the heading Where to Start.

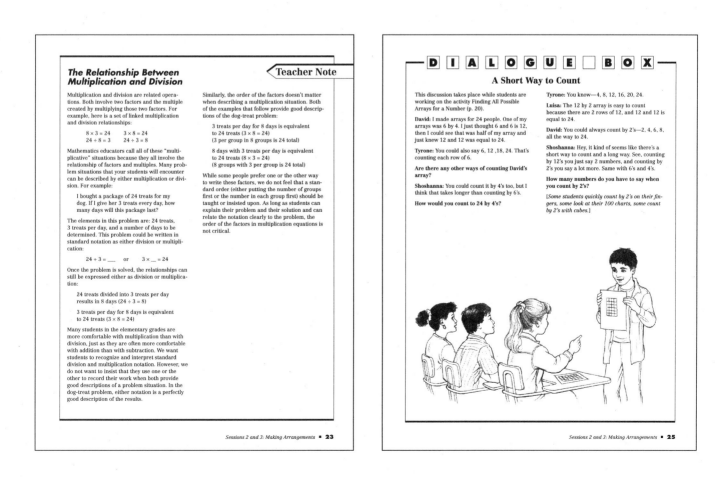

The Relationship Between Multiplication and Division

◁ **Teacher Note**

Multiplication and division are related operations. Both involve two factors and the multiple created by multiplying those two factors. For example, here is a set of linked multiplication and division relationships:

$$8 \times 3 = 24 \qquad 3 \times 8 = 24$$
$$24 \div 8 = 3 \qquad 24 \div 3 = 8$$

Mathematics educators call all of these "multiplicative" situations because they all involve the relationship of factors and multiples. Many problem situations that your students will encounter can be described by either multiplication or division. For example:

I bought a package of 24 treats for my dog. If I give her 3 treats every day, how many days will this package last?

The elements in this problem are: 24 treats, 3 treats per day, and a number of days to be determined. This problem could be written in standard notation as either division or multiplication:

$$24 \div 3 = \_\_ \qquad \text{or} \qquad 3 \times \_\_ = 24$$

Once the problem is solved, the relationships can still be expressed either as division or multiplication:

24 treats divided into 3 treats per day results in 8 days ($24 \div 3 = 8$)

3 treats per day for 8 days is equivalent to 24 treats ($3 \times 8 = 24$)

Many students in the elementary grades are more comfortable with multiplication than with division, just as they are often more comfortable with addition than with subtraction. We want students to recognize and interpret standard division and multiplication notation. However, we do not want to insist that they use one or the other to record their work when both provide good descriptions of a problem situation. In the dog-treat problem, either notation is a perfectly good description of the results.

Similarly, the order of the factors doesn't matter when describing a multiplication situation. Both of the examples that follow provide good descriptions of the dog-treat problem:

3 treats per day for 8 days is equivalent to 24 treats ($3 \times 8 = 24$)
(3 per group in 8 groups is 24 total)

8 days with 3 treats per day is equivalent to 24 treats ($8 \times 3 = 24$)
(8 groups with 3 per group is 24 total)

While some people prefer one or the other way to write these factors, we do not feel that a standard order (either putting the number of groups first or the number in each group first) should be taught or insisted upon. As long as students can explain their problem and their solution and can relate the notation clearly to the problem, the order of the factors in multiplication equations is not critical.

Sessions 2 and 3: Making Arrangements ■ **23**

D I A L O G U E ☐ B O X

A Short Way to Count

This discussion takes place while students are working on the activity Finding All Possible Arrays for a Number (p. 20).

David: I made arrays for 24 people. One of my arrays was 6 by 4. I just thought 6 and 6 is 12, then I could see that was half of my array and just knew 12 and 12 was equal to 24.

Tyrone: You could also say 6, 12 ,18, 24. That's counting each row of 6.

Are there any other ways of counting David's array?

Shoshanna: You could count it by 4's too, but I think that takes longer than counting by 6's.

How would you count to 24 by 4's?

Tyrone: You know—4, 8, 12, 16, 20, 24.

Luisa: The 12 by 2 array is easy to count because there are 2 rows of 12, and 12 and 12 is equal to 24.

David: You could always count by 2's—2, 4, 6, 8, all the way to 24.

Shoshanna: Hey, it kind of seems like there's a short way to count and a long way. See, counting by 12's you just say 2 numbers, and counting by 2's you say a lot more. Same with 6's and 4's.

How many numbers do you have to say when you count by 2's?

[*Some students quickly count by 2's on their fingers, some look at their 100 charts, some count by 2's with cubes.*]

Sessions 2 and 3: Making Arrangements ■ **25**

The *Investigations* curriculum incorporates the use of two forms of technology in the classroom: calculators and computers. Calculators are assumed to be standard classroom materials, available for student use in any unit. Computers are explicitly linked to one or more units at each grade level; they are used with the unit on 2-D geometry at each grade, as well as with some of the units on measuring, data, and changes.

Using Calculators

In this curriculum, calculators are considered tools for doing mathematics, similar to pattern blocks or interlocking cubes. Just as with other tools, students must learn both *how* to use calculators correctly and *when* they are appropriate to use. This knowledge is crucial for daily life, as calculators are now a standard way of handling numerical operations, both at work and at home.

Using a calculator correctly is not a simple task; it depends on a good knowledge of the four operations and of the number system, so that students can select suitable calculations and also determine what a reasonable result would be. These skills are the basis of any work with numbers, whether or not a calculator is involved.

Unfortunately, calculators are often seen as tools to check computations with, as if other methods are somehow more fallible. Students need to understand that any computational method can be used to check any other; it's just as easy to make a mistake on the calculator as it is to make a mistake on paper or with mental arithmetic. Throughout this curriculum, we encourage students to solve computation problems in more than one way in order to double-check their accuracy. We present mental arithmetic, paper-and-pencil computation, and calculators as three possible approaches.

In this curriculum we also recognize that, despite their importance, calculators are not always appropriate in mathematics instruction. Like any tools, calculators are useful for some tasks but not for others. You will need to make decisions about when to allow students access to calculators and when to ask that they solve problems without

them so that they can concentrate on other tools and skills. At times when calculators are or are not appropriate for a particular activity, we make specific recommendations. Help your students develop their own sense of which problems they can tackle with their own reasoning and which ones might be better solved with a combination of their own reasoning and the calculator.

Managing calculators in your classroom so that they are a tool, and not a distraction, requires some planning. When calculators are first introduced, students often want to use them for everything, even problems that can be solved quite simply by other methods. However, once the novelty wears off, students are just as interested in developing their own strategies, especially when these strategies are emphasized and valued in the classroom. Over time, students will come to recognize the ease and value of solving problems mentally, with paper and pencil, or with manipulatives, while also understanding the power of the calculator to facilitate work with larger numbers.

Experience shows that if calculators are available only occasionally, students become excited and distracted when they are permitted to use them. They focus on the tool rather than on the mathematics. In order to learn when calculators are appropriate and when they are not, students must have easy access to them and use them routinely in their work.

If you have a calculator for each student, and if you think your students can accept the responsibility, you might allow them to keep their calculators with the rest of their individual materials, at least for the first few weeks of school. Alternatively, you might store them in boxes on a shelf, number each calculator, and assign a corresponding number to each student. This system can give students a sense of ownership while also helping you keep track of the calculators.

Using Computers

Students can use computers to approach and visualize mathematical situations in new ways. The computer allows students to construct and manipulate geometric shapes, see objects move according to rules they specify, and turn, flip, and repeat a pattern.

This curriculum calls for computers in units where they are a particularly effective tool for learning mathematics content. One unit on 2-D geometry at each of the grades 3–5 includes a core of activities that rely on access to computers, either in the classroom or in a lab. Other units on geometry, measuring, data, and changes include computer activities, but can be taught without them. In these units, however, students' experience is greatly enhanced by computer use.

The following list outlines the recommended use of computers in this curriculum:

Kindergarten
Unit: *Making Shapes and Building Blocks*
  (Exploring Geometry)
Software: *Shapes*
Source: provided with the unit

Grade 1
Unit: *Survey Questions and Secret Rules*
  (Collecting and Sorting Data)
Software: *Tabletop, Jr.*
Source: Broderbund

Unit: *Quilt Squares and Block Towns*
  (2-D and 3-D Geometry)
Software: *Shapes*
Source: provided with the unit

Grade 2
Unit: *Mathematical Thinking at Grade 2*
  (Introduction)
Software: *Shapes*
Source: provided with the unit

Unit: *Shapes, Halves, and Symmetry*
  (Geometry and Fractions)
Software: *Shapes*
Source: provided with the unit

Unit: *How Long? How Far?* (Measuring)
Software: *Geo-Logo*
Source: provided with the unit

Grade 3
Unit: *Flips, Turns, and Area* (2-D Geometry)
Software: *Tumbling Tetrominoes*
Source: provided with the unit

Unit: *Turtle Paths* (2-D Geometry)
Software: *Geo-Logo*
Source: provided with the unit

Grade 4
Unit: *Sunken Ships and Grid Patterns*
  (2-D Geometry)
Software: *Geo-Logo*
Source: provided with the unit

Grade 5
Unit: *Picturing Polygons* (2-D Geometry)
Software: *Geo-Logo*
Source: provided with the unit

Unit: *Patterns of Change* (Tables and Graphs)
Software: *Trips*
Source: provided with the unit

Unit: *Data: Kids, Cats, and Ads* (Statistics)
Software: *Tabletop, Sr.*
Source: Broderbund

The software provided with the *Investigations* units uses the power of the computer to help students explore mathematical ideas and relationships that cannot be explored in the same way with physical materials. With the *Shapes* (grades 1–2) and *Tumbling Tetrominoes* (grade 3) software, students explore symmetry, pattern, rotation and reflection, area, and characteristics of 2-D shapes. With the *Geo-Logo* software (grades 2–5), students investigate rotations and reflections, coordinate geometry, the properties of 2-D shapes, and angles. The *Trips* software (grade 5) is a mathematical exploration of motion in which students run experiments and interpret data presented in graphs and tables.

We suggest that students work in pairs on the computer; this not only maximizes computer resources but also encourages students to consult, monitor, and teach each other. Generally, more than two students at one computer find it difficult to share. Managing access to computers is an issue for every classroom. The curriculum gives you explicit support for setting up a system. The units are structured on the assumption that you have enough computers for half your students to work on the machines in pairs at one time. If you do not have access to that many computers, suggestions are made for structuring class time to use the unit with fewer than five.

Assessment plays a critical role in teaching and learning, and it is an integral part of the *Investigations* curriculum. For a teacher using these units, assessment is an ongoing process. You observe students' discussions and explanations of their strategies on a daily basis and examine their work as it evolves. While students are busy recording and representing their work, working on projects, sharing with partners, and playing mathematical games, you have many opportunities to observe their mathematical thinking. What you learn through observation guides your decisions about how to proceed. In any of the units, you will repeatedly consider questions like these:

- Do students come up with their own strategies for solving problems, or do they expect others to tell them what to do? What do their strategies reveal about their mathematical understanding?

- Do students understand that there are different strategies for solving problems? Do they articulate their strategies and try to understand other students' strategies?

- How effectively do students use materials as tools to help with their mathematical work?

- Do students have effective ideas for keeping track of and recording their work? Do keeping track of and recording their work seem difficult for them?

You will need to develop a comfortable and efficient system for recording and keeping track of your observations. Some teachers keep a clipboard handy and jot notes on a class list or on adhesive labels that are later transferred to student files. Others keep loose-leaf notebooks with a page for each student and make weekly notes about what they have observed in class.

Assessment Tools in the Unit

With the activities in each unit, you will find questions to guide your thinking while observing the students at work. You will also find two built-in assessment tools: Teacher Checkpoints and embedded Assessment activities.

Teacher Checkpoints The designated Teacher Checkpoints in each unit offer a time to "check in" with individual students, watch them at work, and ask questions that illuminate how they are thinking.

At first it may be hard to know what to look for, hard to know what kinds of questions to ask. Students may be reluctant to talk; they may not be accustomed to having the teacher ask them about their work, or they may not know how to explain their thinking. Two important ingredients of this process are asking students open-ended questions about their work and showing genuine interest in how they are approaching the task. When students see that you are interested in their thinking and are counting on them to come up with their own ways of solving problems, they may surprise you with the depth of their understanding.

Teacher Checkpoints also give you the chance to pause in the teaching sequence and reflect on how your class is doing overall. Think about whether you need to adjust your pacing: Are most students fluent with strategies for solving a particular kind of problem? Are they just starting to formulate good strategies? Or are they still struggling with how to start? Depending on what you see as the students work, you may want to spend more time on similar problems, change some of the problems to use smaller numbers, move quickly to more challenging material, modify subsequent activities for some students, work on particular ideas with a small group, or pair students who have good strategies with those who are having more difficulty.

Embedded Assessment Activities Assessment activities embedded in each unit will help you examine specific pieces of student work, figure out what they mean, and provide feedback. From the students' point of view, these assessment activities are no different from any others. Each is a learning experience in and of itself, as well as an opportunity for you to gather evidence about students' mathematical understanding.

The embedded assessment activities sometimes involve writing and reflecting; at other times, a discussion or brief interaction between student and teacher; and in still other instances, the creation and explanation of a product. In most cases, the assessments require that students *show* what they did, *write* or *talk* about it, or do both. Having to explain how they worked through a problem helps students be more focused and clear in their mathematical thinking. It also helps them realize that doing mathematics is a process that may involve tentative starts, revising one's approach, taking different paths, and working through ideas.

Teachers often find the hardest part of assessment to be interpreting their students' work. We provide guidelines to help with that interpretation. If you have used a process approach to teaching writing, the assessment in *Investigations* will seem familiar. For many of the assessment activities, a Teacher Note provides examples of student work and a commentary on what it indicates about student thinking.

Documentation of Student Growth

To form an overall picture of mathematical progress, it is important to document each student's work. Many teachers have students keep their work in folders, notebooks, or journals, and some like to have students summarize their learning in journals at the end of each unit. It's important to document students' progress, and we recommend that you keep a portfolio of selected work for each student, unit by unit, for the entire year. The final activity in each *Investigations* unit, called Choosing Student Work to Save, helps you and the students select representative samples for a record of their work.

This kind of regular documentation helps you synthesize information about each student as a mathematical learner. From different pieces of evidence, you can put together the big picture. This synthesis will be invaluable in thinking about where to go next with a particular child, deciding where more work is needed, or explaining to parents (or other teachers) how a child is doing.

If you use portfolios, you need to collect a good balance of work, yet avoid being swamped with an overwhelming amount of paper. Following are some tips for effective portfolios:

- Collect a representative sample of work, including some pieces that students themselves select for inclusion in the portfolio. There should be just a few pieces for each unit, showing different kinds of work—some assignments that involve writing as well as some that do not.

- If students do not date their work, do so yourself so that you can reconstruct the order in which pieces were done.

- Include your reflections on the work. When you are looking back over the whole year, such comments are reminders of what seemed especially interesting about a particular piece; they can also be helpful to other teachers and to parents. Older students should be encouraged to write their own reflections about their work.

Assessment Overview

There are two places to turn for a preview of the assessment opportunities in each *Investigations* unit. The Assessment Resources column in the unit Overview Chart identifies the Teacher Checkpoints and Assessment activities embedded in each investigation, guidelines for observing the students that appear within classroom activities, and any Teacher Notes and Dialogue Boxes that explain what to look for and what types of student responses you might expect to see in your classroom. Additionally, the section About the Assessment in This Unit gives you a detailed list of questions for each investigation, keyed to the mathematical emphases, to help you observe student growth.

Depending on your situation, you may want to provide additional assessment opportunities. Most of the investigations lend themselves to more frequent assessment, simply by having students do more writing and recording while they are working.

Arrays and Shares

Content of This Unit   Through the work in this unit—skip counting, building arrays, dividing quantities into equal shares, and solving word problems—students develop a clear sense of what multiplication and division are and how these processes are related. They use 100 charts, rectangular arrays, games, and cubes to gain fluency with multiplication and division pairs and to learn how to solve multiplication and division problems by breaking them into manageable components. Students are encouraged to develop their own strategies, based on what they know about numbers, for solving multiplication and division problems.

Students are asked to analyze real-world applications of multiplication and division situations and represent them mathematically to find solutions; they also create their own real-world problems for mathematical statements.

Connections with Other Units   If you are doing the full-year *Investigations* curriculum in the suggested sequence for Grade 4, this is the second of eleven units. Your class will already have had some experience with skip counting on the 100 chart in the introductory unit, *Mathematical Thinking at Grade 4*. There, students looked for patterns on the 100 chart as they skip counted.

The work in this unit is also continued and extended in the Grade 4 units *Landmarks in the Thousands*, using larger numbers with factors and multiples of 100, and *Packages and Groups*, finding factors of larger numbers and solving multidigit multiplication and division problems.

If your school is not using the full-year curriculum, this unit can also be used successfully in Grade 5.

Investigations Curriculum ■ Suggested Grade 4 Sequence

Mathematical Thinking at Grade 4 (Introduction)

▶ *Arrays and Shares* (Multiplication and Division)

Seeing Solids and Silhouettes (3-D Geometry)

Landmarks in the Thousands (The Number System)

Different Shapes, Equal Pieces (Fractions and Area)

The Shape of the Data (Statistics)

Money, Miles, and Large Numbers (Addition and Subtraction)

Changes Over Time (Graphs)

Packages and Groups (Multiplication and Division)

Sunken Ships and Grid Patterns (2-D Geometry)

Three out of Four Like Spaghetti (Data and Fractions)

Investigation 1 ■ Multiples on the 100 Chart

Class Sessions	Activities	Pacing
Sessions 1 and 2 (p. 4) MULTIPLES ON THE 100 CHART	Skip Counting on the 100 Chart Highlighting 100 Charts Homework: Multiples on the 100 Chart Extension: Divisibility Rules	minimum 2 hr
Session 3 (p. 8) SKIP COUNTING AND MULTIPLYING	Making Groups Solving Multiplication Clusters Teacher Checkpoint: Multiplication Clusters Homework: More Multiplication Clusters	minimum 1 hr

Mathematical Emphasis

- Using skip counting as a model for multiplication

- Seeing multiplication as an accumulation of groups of a number

- Looking for the multiplication patterns of numbers

- Interpreting standard multiplication and division notation

- Using known multiplication relationships to solve harder relationships

Assessment Resources

Introducing Mathematical Vocabulary (Teacher Note, p. 6)

Students' Problems with Skip Counting (Teacher Note, p. 7)

Teacher Checkpoint: Multiplication Clusters (p. 10)

What About Notation? (Teacher Note, p. 11)

Materials

Markers or crayons

Interlocking cubes

Overhead projector

Student Sheets 1–3

Family letter

Teaching resource sheets

Stapler, staples

Investigation 2 ▪ Arrays

Class Sessions	Activities	Pacing
Session 1 (p. 14) THINGS THAT COME IN ARRAYS	Finding Arrays in and out of Our Classroom Homework: Things That Come in Arrays	minimum 1 hr
Sessions 2 and 3 (p. 18) MAKING ARRANGEMENTS	Discussing Homework Arranging a Group of 18 People Finding All Possible Arrays for a Number Class Discussion: Looking at All the Arrays Homework: Arranging Chairs Extension: Arrays for Higher Numbers Extension: Prime Numbers	minimum 2 hr
Session 4 (p. 26) PREPARING A SET OF ARRAYS	Preparing a Personal Set of Array Cards Using the Array Cards	minimum 1 hr
Sessions 5 and 6 (p. 29) ARRAY GAMES	Another Array Game Choice Time: Multiplying, Dividing, and Skip Counting Teacher Checkpoint: Choice Time How Many Ways Can You Make 8×6? Homework: Pairs I Know and Pairs I Don't Know Homework: Array Games	minimum 2 hr
Sessions 7 and 8 (p. 35) LOOKING AT DIVISION	Introducing Division Notation Division Problems with Leftovers Class Discussion: Division with "Extras" Assessment: Creating and Solving Division Problems Homework: Word Problems	minimum 2 hr

🕐 Ten-Minute Math ▪ Counting Around the Class

Mathematical Emphasis

- Using an array as a model for multiplication

- Becoming more familiar with multiplication pairs

- Recognizing prime numbers as those that each have only one pair of factors and only one array

- Becoming comfortable with a variety of notation used for multiplication and division

- Understanding how division notation represents a variety of division situations

- Determining what to do with "leftovers" in division, depending on the situation

Assessment Resources

The Relationship Between Multiplication and Division (Teacher Note, p. 23)

Things That Come in Arrays (Dialogue Box, p. 24)

A Short Way to Count (Dialogue Box, p. 25)

Array Games (Teacher Note, p. 28)

Teacher Checkpoint: Choice Time (p. 32)

Assessment: Creating and Solving Division Problems (p. 38)

Two Kinds of Division: Sharing and Partitioning (Teacher Note, p. 39)

Talking and Writing About Division (Teacher Note, p. 41)

Materials

Interlocking cubes

Construction paper

Scissors, glue, paste, tape

Quart-size resealable bags

Calculators

Overhead projector

Student Sheets 4–9

Teaching resource sheets

Chart paper

Objects arranged in arrays

Array Cards

Investigation 3 ▪ Multiplication and Division with Two-Digit Numbers

Class Sessions	Activities	Pacing
Session 1 (p. 44) MULTIPLICATION CLUSTERS	Solving Cluster Problems Writing About Your Strategies Homework: Another Set of Related Problems	minimum 1 hr
Sessions 2, 3, and 4 (p. 49) MULTIPLICATION AND DIVISION CHOICES	Choice Time: Multiplication and Division Activities Class Discussion: Strategies for Solving Two-Digit Multiplication Problems Class Discussion: How Many Boxes Do You Need? Homework: Multiplication Clusters at Home Homework: More Array Games	minimum 3 hr
Session 5 (p. 55) PROBLEMS THAT LOOK HARD BUT AREN'T	Discussing Multiplication Pairs Which Pairs Are Hard for You? Problems That Look Hard But Really Aren't Assessment: Solving Two-Digit Multiplication Problems Choosing Student Work to Save Homework: Problems That Look Hard But Aren't	minimum 1 hr

◕ Ten-Minute Math ▪ Multiple BINGO

Mathematical Emphasis

- Becoming fluent in basic multiplication relationships

- Partitioning numbers to multiply them more easily

- Recognizing multiplication and division situations and representing each situation using a mathematical statement

- Learning about patterns that are useful for multiplying by multiples of 10

Assessment Resources

Cluster Problems (Teacher Note, p. 47)

Ways to Solve 3 × 24 (Dialogue Box, p. 48)

Multiplying by Multiples of 10 (Teacher Note, p. 54)

Assessment: Solving Two-Digit Multiplication Problems (p. 56)

Choosing Student Work to Save (p. 57)

Two Ways to Solve 27 × 4 (Teacher Note, p. 58)

Strategies for Learning "Hard Problems" (Dialogue Box, p. 59)

Materials

Interlocking cubes

Overhead projector, transparency pen

Student Sheets 10–16

Teaching resource sheets

Array Cards

Following are the basic materials needed for the activities in this unit. Many of the items can be purchased from the publisher, either individually or in the Teacher Resource Package and the Student Materials Kit for grade 4. Detailed information is available on the *Investigations* order form. To obtain this form, call toll-free 1-800-872-1100 and ask for a Dale Seymour customer service representative.

Snap™ Cubes (interlocking cubes): at least 50 per student pair

Array Cards, Sets A and B (manufactured), or use blackline masters to make your own sets: 1 set per pair (class); 1 set per student (homework)

Calculators: at least 1 per pair

Chart paper: 1 sheet per student

Construction paper (12" by 18"): 1 per student

Quart-size resealable plastic bags: 1 per student

Scissors, glue or paste, tape

Overhead projector, transparency pen

Markers or crayons

Stapler, staples

Objects arranged in arrays

The following materials are provided at the end of this unit as blackline masters. A Student Activity Booklet containing all student sheets and teaching resources needed for individual work is available.

Related Children's Literature

Giganti, Paul Jr. *Each Orange Had 8 Slices.* New York: Greenwillow Books, 1992.

Hong, Lily Toy. *Two of Everything.* Morton Grove, Ill.: Albert Whitman and Co., 1993

Schwartz, Amy. *Annabelle Swift, Kindergartner.* New York: Orchard Books, 1988.

This unit focuses on the process and application of multiplication and division. The multiplication and division situations in this unit deal with equal groups. Multiplication is used when the size of the group and the number of groups are known and we want to find the total number of items. Division is most often used when the total quantity is known and we are trying to find out either the number of groups or the size of the groups.

Traditional instruction in these important arithmetic processes emphasizes speed and memorization of single-digit "facts" and computation procedures. However, students may know their "facts" without being able to recognize situations in which multiplication and division are useful. They may learn procedures without having a sense of how multiplication and division really work.

The focus of this unit is on supporting students to make sense out of multiplication and division. As students develop strategies to use in multiplication and division situations, it is critical that they develop visual images that support their work. Through using an array model (a rectangular arrangement of objects in rows and columns), they begin to visualize important multiplication relationships, for example, that the solution to 7×6 is the same as the solution to 6×7. By combining arrays into larger arrays, they develop strategies to solve two-digit multiplication and related division problems. For example, the problem 7×23 can be solved by breaking the problem into more familiar parts: 7×10, 7×10, and 7×3.

Students need to develop efficient computation strategies, many of which will be mental strategies, but these must be based on their understanding of the quantities and their relationships, not on memorized procedures. Good mental strategies often start from the left, focusing first on the largest part of the number, rather than the smallest: For 56×8, we might think, "eight 50's is equal to 400, eight 6's is equal to 48, so that's 448." When students see standard notation for multiplication or division such as

$$\begin{array}{r} 56 \\ \times\, 8 \end{array} \qquad 8\,\overline{)\,432}$$

the form of the problem may trigger use of poorly understood, and often inefficient, algorithms. For example, in the first problem, students might start to say, "8 times 6 is equal to 48, put down the 8 and carry the 4." This procedure obscures the use of good number sense and often leads students to fragment a number into its digits and lose track of the quantities represented by the numerals. We would like students to recognize and interpret all the standard notations they are likely to see in elementary school but to continue to solve problems using good number sense and good strategies built on an understanding of the number relationships involved in the problem.

At the beginning of each investigation, the Mathematical Emphasis section tells you what is most important for students to learn about during that investigation. Many of these mathematical understandings and processes are difficult and complex. Students gradually learn more and more about each idea over many years of schooling. Individual students will begin and end the unit with different levels of knowledge and skill, but all will gain greater understanding of multiplication and division situations and strategies for solving these problems based on good number sense and number relationships.

Throughout the *Investigations* curriculum, there are many opportunities for ongoing daily assessment as you observe, listen to, and interact with students at work. In this unit, you will find two Teacher Checkpoints:

Investigation 1, Session 3:
Multiplication Clusters (p. 10)

Investigation 2, Sessions 5–6:
Choice Time (p. 32)

This unit also has two embedded assessment activities:

Investigation 2, Sessions 7–8:
Creating and Solving Division Problems (p. 38)

Investigation 3, Session 5:
Solving Two-Digit Multiplication Problems (p. 56)

In addition, you can use almost any activity in this unit to assess your students' needs and strengths. Listed below are questions to help you focus your observation in each investigation. You may want to keep track of your observations for each student to help you plan your curriculum and monitor students' growth. Suggestions for documenting student growth can be found in the section About Assessment.

Investigation 1: Multiples on the 100 Chart

- How comfortable and accurate are students with skip counting? How do they keep track of their work?

- How do students visualize and view multiplication? Do they demonstrate an understanding of multiplication as the accumulation of equal groups?

- What do students notice about patterns of multiples highlighted on the 100 chart?

- How do students recognize, interpret, and use standard notation for multiplication and division?

- How do students solve multiplication problems? Do they use what they already know about landmarks in the number system and other familiar number relationships?

Investigation 2: Arrays

- How do students view and use arrays? Do they connect arrays with multiplication? How do students count their arrays?

- How familiar are students with multiplication pairs? What strategies do students use to solve multiplication pairs? Are students able to find combinations of smaller arrays that equal a bigger array?

- How do students find and describe prime numbers? Do they recognize that prime numbers have only one pair of factors and one array?

- How comfortable are students with various multiplication and division symbols and notations? Do they choose an appropriate strategy for the problem regardless of symbols or notation?

- How do students interpret problems presented with division notation? Do they understand that division notation can represent a variety of division situations?

- How do students handle leftovers or remainders? Do their strategies make sense given the content of the problem?

Investigation 3: Multiplication and Division with Two-Digit Numbers

- How familiar are students with multiplication pairs and relationships? What strategies do students use to solve multiplication pairs?

- What strategies do students use to solve cluster problems? Do they partition large numbers into more familiar parts?

- How do students recognize and interpret multiplication and division situations? Are they able to write an equation that matches the situation?

- How do students use patterns to solve multiplication and division problems? Do they use patterns to multiply by multiples of 10?

In the *Investigations* curriculum, mathematical vocabulary is introduced naturally during the activities. We don't ask students to learn definitions of new terms; rather, they come to understand such words as *factor* or *area* or *symmetry* by hearing them used frequently in discussion as they investigate new concepts. This approach is compatible with current theories of second-language acquisition, which emphasize the use of new vocabulary in meaningful contexts while students are actively involved with objects, pictures, and physical movement.

Listed below are some key words used in this unit that will not be new to most English speakers at this age level but may be unfamiliar to students with limited English proficiency. You will want to spend additional time working on these words with your students who are learning English. If your students are working with a second-language teacher, you might enlist your colleague's aid in familiarizing students with these words, before and during this unit. In the classroom, look for opportunities for students to hear and use these words. Activities you can use to present the words are given in the appendix, Vocabulary Support for Second-Language Learners (p. 64).

packages, packaged   Students examine items such as juice boxes and pencils that are packaged or arranged in rectangular arrays. They look at the dimensions of these packages as one way of determining the total number of items. For example, eggs are usually packaged in 2 rows of 6 (2×6). Students could skip count by 2's as a way of determining the total number of eggs, or they might add $6 + 6$ to figure the total.

related   Students solve groups or clusters of multiplication problems that are related to each other. Some problems in the cluster might be related by doubles (4×5 and 4×10), or they might be multiples of 10 (3×2 and 3×20). Other problems might be broken down into smaller, more familiar problems, such as 4×5, 4×10, and 4×15.

strategy, strategies   Throughout the unit, students are encouraged to share their ideas and approaches to solving problems. The emphasis is on solving problems and using ideas, processes, and strategies that make sense to the student. Students are encouraged to share their problem-solving strategies with their classmates. By doing this, students experience that there are many ways to solve a problem.

Multicultural Extensions for All Students

Whenever possible, encourage students to share words, objects, customs, or any aspects of daily life from their own cultures and backgrounds that are relevant to the activities in this unit. For example:

■ When students are making up problem situations that represent division situations, encourage them to write problems that are based on aspects of their cultures—foods, games, objects, and sports that involve teams, and so on.

Investigations

Multiples on the 100 Chart

What Happens

Sessions 1 and 2: Multiples on the 100 Chart
Students are introduced to skip counting on the 100 chart. They create a set of highlighted 100 charts that show the multiplication patterns of numbers. Students use these charts throughout the unit to practice skip counting and play games.

Session 3: Skip Counting and Multiplying
Students use interlocking cubes and skip counting on the 100 chart as tools for determining how many groups of a smaller number are in a larger number. They discuss how these problems can be represented using standard multiplication and division notation. They solve clusters of multiplication problems, using one solution to help them find others.

Mathematical Emphasis

■ Using skip counting as a model for multiplication (for example, count by 2's to see that six 2's is equal to 12)

■ Seeing multiplication as an accumulation of groups of a number (for example, use cubes or the 100 chart to show that 6 groups of 3 is equal to 18)

■ Looking for the multiplication patterns of numbers (including patterns of multiples highlighted on the 100 chart)

■ Interpreting standard multiplication and division notation

■ Using known multiplication relationships to solve harder relationships (for example, three 6's is equal to 18, so six 6's must be equal to 36 because 18 and 18 is equal to 36)

What to Plan Ahead of Time

Materials

- Markers or crayons (Sessions 1–2)
- Interlocking cubes: at least 60 per student (Session 3)
- Overhead projector, transparency pen (Sessions 1–2)
- Stapler, staples

Other Preparation

- Duplicate student sheets and teaching resources (located at the end of this unit) in the following quantities. If you have Student Activity Booklets, copy only the item marked with an asterisk.

  For Sessions 1–2

  100 chart (p. 72): 1 per student

  100 chart (4 to a page) (p. 73): 3 per student

  Family letter* (p. 68): 1 per student. Remember to sign it before copying.

  Student Sheet 1, Multiples on the 100 Chart (p. 69): 1 per student (homework)

  For Session 3

  Student Sheet 2, Multiplication Clusters (p. 70): 1 per student

  Student Sheet 3, More Multiplication Clusters (p. 71): 1 per student (homework)

  100 chart (p. 72): 1 per student (class), 1 per student (homework, optional)

- Assemble (or have the students assemble) interlocking cubes in towers of 10. Ask students to return cubes in groups of ten when they are finished using them. (Session 3)

- Make an overhead transparency of the 100 chart (p. 72). (Sessions 1–2)

- If you plan to provide folders in which students will save their work for the entire unit, prepare these for distribution during Session 1.

- If you do not have manufactured Array Cards for Investigations grade 4, you may want to use the blackline masters in the back of this book to make sets of Array Cards for Investigation 2. See p. 82 for Array Cards Set A (sheets 1–6) and p. 88 for Array Cards Set B (sheets 7–17).

  Read the discussion on p. 26 to determine the number of sets your classroom aides or parent volunteers will need to cut out and, in some cases, label. You may want to give your aides copies of How to Make Array Cards (p. 81) to guide them in cutting out and labeling the cards.

  Each set of Array Cards, manufactured or not, should be stored in a quart-size resealable plastic bag with the name or names of the appropriate student or student pairs written on it in permanent marker.

Materials

- 100 chart (1 per student)
- 100 chart (4 to a page) (3 per student)
- Markers or crayons
- Transparency of 100 chart
- Overhead projector, transparency pen
- Family letter (1 per student)
- Student Sheet 1 (1 per student, homework)
- Stapler, staples

Multiples on the 100 Chart

What Happens

Students are introduced to skip counting on the 100 chart. They create a set of highlighted 100 charts that show the multiplication patterns of numbers. Students use these charts throughout the unit to practice skip counting and play games. Their work focuses on:

- practicing skip counting
- looking for the multiplication patterns of numbers 2 to 12

Activity

Skip Counting on the 100 Chart

Put a 100 chart on the overhead. Ask students which numbers would get circled if you were counting by 2's. Circle these numbers on the 100 chart. This is an appropriate time to integrate *multiple* as a word to describe the numbers you are circling. See the **Teacher Note**, Introducing Mathematical Vocabulary (p. 6).

It is not necessary to circle the numbers in order. In fact, skip around a bit—asking for some suggestions of multiples of 2 in the 40's or 70's. When you have a few multiples circled, ask some students or the whole class to count slowly in unison by 2's while you circle all the multiples of 2.

Finding Patterns When all the multiples of 2 are circled on the chart, have students describe the patterns they see for the 2's. Write their observations below the chart.

Pass out 100 charts for students to circle or highlight the multiples of 2. In addition to circling or highlighting on their charts, they should write one statement about a pattern they notice. As students finish, they can practice skip counting by 2's with a partner, using their marked 100 chart if they need it. One way to do this is for one student to look at the chart while the other counts without seeing the chart; the first student checks the count and provides hints as necessary.

❖ **Tip for the Linguistically Diverse Classroom** Have limited-English-proficient students highlight a pattern they recognize on the circled chart (or circle the pattern if they've highlighted the numbers).

Highlighting 100 Charts

Give each student three copies of the 100 chart (4 to a page). Each student makes circled or highlighted 100 charts for the multiples of 2 to 12. Although each student makes his or her own set of charts, students should work in pairs or small groups.

As you circulate around the room, encourage students to check with one another continually; if they find differences, they need to figure out what to correct. Also, encourage conversations within groups about what patterns they see and what shortcuts they find in order to complete their charts.

At the bottom of each chart students should write a statement about the pattern they notice on the marked 100 chart.

❖ **Tip for the Linguistically Diverse Classroom**   Have limited-English-proficient students highlight or circle a pattern they recognize for each 100 chart of multiples 2 to 12.

When students have finished a chart, they should use their chart and practice skip counting by that number with a partner as a way of familiarizing themselves with the number sequence and checking the pattern on their 100 chart. Skip counting can be confusing for some students, and it is also quite easy to make an error. See the **Teacher Note**, Students' Problems with Skip Counting (p. 7), for some issues to watch out for and ideas about how to support students who might have difficulty.

Sharing Patterns   Pose some questions about the group of charts for the class to investigate:

Are there charts that have only even numbers marked?

Are there charts that have only odd numbers marked?

What numbers never seem to be marked? Why do you think that is true? On what other charts might they be marked?

What else did you notice?

Students can staple their pages of 100 charts together as a booklet and keep them in their math folders to refer to as needed. As they become more familiar with the patterns of the multiples of 2 to 12, they will probably use their charts less and less. Since these charts are useful for reference throughout the year, some teachers have had their students make them into more permanent books with sturdy covers.

Sessions 1 and 2 Follow-Up

Homework

Multiples on the 100 Chart   Send home the family letter or the *Investigations* at Home booklet and Student Sheet 1, Multiples on the 100 Chart. Students complete the four charts for multiples of numbers greater than 12, such as 15 and 16. (If students have not finished their booklet of charts for 2–12, they may work on those charts in addition to or instead of Student Sheet 1.) Remind students to write one statement about a pattern they notice for each completed chart. If possible, they can practice skip counting with an adult or sibling.

Extension

Divisibility Rules   Some students may be interested in finding "divisibility rules." They might make posters as they work out rules for individual numbers.

Teacher Note ⟩ *Introducing Mathematical Vocabulary*

This unit provides the opportunity for introducing several important mathematical words naturally. Introduce these words by beginning to use them yourself and explaining what you mean by them, but don't insist that students use them. If the introduction of a vocabulary word is preceded by activities that make its definition clear, students enjoy knowing an "adult" word to refer to a new concept they have learned.

Multiple   *Multiple* is a natural word to introduce to students as they count by 2's (or any other number) and mark those numbers on the 100 chart. During the activity refer to the numbers you have marked as "the multiples of 2." You might ask students if they have other names for these numbers or for the counting, such as "the 2's table" or "counting by 2's."

Factor   A *factor* of a number is a number that can be divided evenly into another number. For example, 2 is a factor of 4, 6, 8, and all the even numbers; 1, 2, 4, 8, and 16 are all factors of 16. The idea of factors comes up naturally when students make and use arrays. The dimensions of each array are factors of the total number in the array.

Even and Odd   These words will come up in the students' descriptions of patterns on the 100 chart. Don't assume that students know exactly what they mean by these words. Some children believe that an even number has only even factors: "No, 3 isn't a factor of 24 because 3 isn't even." This is a good conjecture to have students investigate.

Row and Column   When talking about their ways of working with the 100 chart, students often confuse the words *row* and *column,* describing a pattern as going "down the row" rather than "down the column." This may be a good opportunity to talk about the difference, since using *row* for both (as students often do) makes communication more difficult. However, do not insist that students use these words in the conventional way as long as they can explain or demonstrate what they mean. Remembering the difference can be honestly confusing, and focusing on getting the words right may obscure the good mathematical thinking a student is doing. Rather, keep using them yourself so students continually hear them used correctly in context. Other terms that may come up in this context and may need some explanation are *horizontal,* *vertical,* and *diagonal.*

Students' Problems with Skip Counting

Some students have difficulty keeping track of their count on the 100 chart. Here are some confusions we have noticed in classrooms:

- Students sometimes start on 1, no matter which number they are skip counting by.

- Students' count is off by 1 because they start counting with the last number they circled or highlighted. For example, when counting by 6's, a student counts 6, 12, 18, and then begins his or her next count on 18, counts six numbers (18, 19, 20, 21, 22, 23), and lands on 23 instead of 24.

- Students follow a "false pattern" that doesn't actually work for the number they are counting by. For example, some students circle or highlight 3, 6, 9, and then mark straight down the columns under the 3, 6, and 9, not realizing the 3's pattern doesn't continue in columns as the 2's pattern does.

- Students miscount one interval and then continue counting correctly so that all the subsequent numbers are affected by the original mistake—for example: 3, 6, 9, 12, 15, 19, 22, 25, 28, and so on.

Some of these difficulties are simply the kind of miscounting mistakes anyone can make. Help students use the pattern on their 100 charts to check: Does the pattern continue consistently on the chart? Also, have students double-check one another. When two or three students compare charts, they can often find and correct their own miscounting.

However, some students may truly not understand what they are doing when they "count by 2's" or "count by 3's" on their charts. These students can build their 2's or 3's with cubes first. A student makes a group of 2 cubes, then marks 2 on the 100 chart; then makes another group of 2 cubes (perhaps in a different color), and then marks the total, 4; then makes another group of cubes, marks the total 6; and so forth. Allow students to stop using the cubes as soon as they feel comfortable.

We have found that it is not helpful for students to use cubes to mark squares directly on the 100 charts. Students can't see the numbers underneath them, and they often move a cube accidentally to a neighboring square, thereby misleading themselves about the pattern on the chart.

Skip Counting and Multiplying

Materials

- Interlocking cubes in towers of 10 (at least 60 per student)
- 100 chart (1 per student, class; 1 per student, homework, optional)
- Student Sheet 2 (1 per student)
- Student Sheet 3 (1 per student, homework)

What Happens

Students use interlocking cubes and skip counting on the 100 chart as tools for determining how many groups of a smaller number are in a larger number. They discuss how these problems can be represented using standard multiplication and division notation. They solve clusters of multiplication problems, using one solution to help them find others. Their work focuses on:

- representing multiplication
- interpreting multiplication and division notation
- solving related multiplication problems
- skip counting

Activity

Making Groups

Make sure all students have access to interlocking cubes. Pose the following question to students as a way of connecting the idea of "counting by 2's" to accumulating groups of 2's.

Using your cubes, show how many 2's it would take to make 24.

Circulate quickly as students are building to check that they can demonstrate that it takes 12 groups of 2 to make 24.

How could you show the same thing on your 100 chart? Is there a way to use the chart to show that 12 groups of 2 make 24?

Take time to explore this question. You may need to follow up with questions, such as:

How many groups of 2 have you counted when you skip count to 4? to 12? How can you tell from the chart?

Pose another, related problem:

How many 2's are in 48?

After students talk with a partner about this problem for a minute or two, ask them to present their solutions. Students might solve this problem using cubes, using the 100 chart, or relating this problem to the previous one.

Using Standard Multiplication and Division Notation   Ask students how they would write down "24 groups of two make 48" with numbers or with words. Record their suggestions:

24 groups of 2 make 48 $2 + 2 + 2 + 2 + ... = 48$

$$24 \times 2 = 48 \qquad \begin{array}{r} 24 \\ \times\, 2 \\ \hline 48 \end{array}$$

If no one suggests the standard multiplication notation, $24 \times 2 = 48$, introduce it yourself and read it as "24 groups of 2 is equal to 48." Show students that it can be written vertically or horizontally.

Note: At this point, you might want to introduce or students might bring up the related division problems:

$48 \div 2$ $\qquad 2\overline{)48}$ $\qquad 48 \div 24$ $\qquad 24\overline{)48}$

Students might read the first two problems as "How many 2's are in 48?" or "Divide 48 into 2 parts." We will focus on interpreting division notation in Investigation 2.

Solving Multiplication Clusters

Multiplication Clusters are sets of problems that help students think about how to use multiplication relationships they know to build solutions to harder multiplication problems. Write the following cluster on the board:

$$3 \times 3 \qquad\qquad 6 \times 3 \qquad\qquad 12 \times 3$$

We're going to solve 12×3. Some of you might know the solution to 12×3 and some of you may not. But let's say we all don't know. How can knowing that 3×3 is equal to 9 help you solve 6×3? How does the solution to 6×3 help you with 12×3? If you have 6 groups of 3 cubes in front of you, how many more groups will you have to make in order to have 12 groups? How many cubes would that be? Can anyone think of any other problem you already know that might help you solve 12×3 in a different way?

Don't worry if not all students are able to use these strategies right away; they will have many more opportunities in this unit to work on these kinds of relationships.

Tell students they will be working on "clusters" of multiplication problems they can solve using their 100 charts or cubes. Refer to any strategies students discovered for the previous problem. For more about cluster problems, see the **Teacher Note**, Cluster Problems (p. 47).

Teacher Checkpoint

Multiplication Clusters

Working in pairs, each student completes a copy of Student Sheet 2, Multiplication Clusters. As you circulate, make sure students can prove their solutions using the 100 charts or cubes and can interpret the multiplication problems meaningfully. If students say, "It's 48 because 4 times 2 is equal to 8 and 4 times 1 is equal to 4" for 12×4, it is likely that they are only seeing individual digits. Your role in supporting students to interpret written notation meaningfully is critical (see the **Teacher Note**, What About Notation?, p. 11). Particularly notice if students are using the first two problems in the cluster to solve the last problem. The following questions will help you think about your students' work.

- Are students reasoning about how one multiplication expression relates to another (for example, that 12×4 is double 6×4)?

- Are students able to split a multiplication problem into parts, solve each part, then put the parts together to give a complete solution (for example, 7×6 is equal to 5×6 plus 2×6, so $30 + 12$ is 42)?

- Are students coming up with their own ideas about how they can use a problem they know to solve a problem they don't know?

Session 3 Follow-Up

More Multiplication Clusters Send home Student Sheet 3, More Multiplication Clusters, for homework. Students can also take home a fresh 100 chart to help them skip count and to doublecheck their solutions.

What About Notation?

It is important that your students learn to recognize, interpret, and use the standard forms and symbols for multiplication and division, both on paper and on the calculator. In this unit, students will use these:

$$\begin{array}{c} 12 \\ \underline{\times\ 3} \end{array} \qquad 3 \times 12 \qquad 12 \div 3 \qquad 3\overline{)12}$$

Your challenge is to introduce these symbols in a way that allows students to interpret them meaningfully. That is, students must understand what is being asked in a problem that is written in standard notation. They can then devise their own way to find an answer. Notation is also useful as an efficient way to record a problem and its solution. It is not just a directive to carry out a particular procedure, or a signal to forget everything you ever knew about the relationships of the numbers in the problem!

Your students may come to you already believing that when they see a problem like $3\overline{)42}$, written in the familiar division format, they must carry out the traditional long-division procedure. Instead, we want them to use everything they know about these two numbers in order to solve the problem. They might skip count by 3's out loud or on the calculator. (For tips on skip counting on the calculator, see the Ten-Minute Math activity, p. 61.) Or they might use reasoning based on their understanding of number relationships:

It takes ten 3's to make 30. Then there are three more 3's to get up to 39, that's thirteen 3's so far. Then 40, 41, 42—that's one more 3—it's 14!

Well, half of 42 is 21, and I can divide 21 into seven groups of 3, so you double that, and it's 14.

Similarly, when students see a multiplication problem like 4×55 written vertically, they are likely to forget everything they know about these numbers and try to carry out multiplication with carrying. Instead, we want students to use what they know about landmarks in the number system and other familiar number relationships. For example:

I know that two 50's make 100, and there's four 50's, so that's 200. Then I know that four 5's is 20, so it's 220.

Students need to get used to interpreting multiplication in both horizontal and vertical form as simply indicating a multiplication situation, not a particular way to carry out the problem. So, while you help students to read standard notation and to use it to record their work, keep the emphasis on understanding the problem context and using good number sense to solve the problem.

INVESTIGATION 2

Arrays

What Happens

Session 1: Things That Come in Arrays
Students think of items that are packaged or arranged in arrays (a rectangular arrangement of objects in rows and columns), such as juice cans or window panes. They record these types of arrays, their dimensions, and the total number of items.

Sessions 2 and 3: Making Arrangements Using cubes and graph paper, students make all the possible arrays for a number. They make a poster of all the arrays, labeling the dimensions of each. Students discuss what they have observed about the class set of posters and explore prime and composite numbers.

Session 4: Preparing a Set of Arrays Students prepare a personal set of Array Cards for use in homework assignments throughout the remainder of the unit. They label the Array Cards in Set A, consisting of 51 arrays, which represent the multiplication combinations of the factors 2 through 12 with totals (or products) up to 50. Later in the unit, or as they become fluent working with these cards, students gradually add the larger arrays in Set B. Students begin to learn several games that use the Array Cards.

Sessions 5 and 6: Array Games During these two sessions, students are introduced to Choice Time. They have the opportunity to play three array games and to practice skip counting in pairs or small groups. As they become more comfortable and familiar with these arrays, they add larger numbers to their sets until the sets represent all multiplication combinations with factors 2 through 12.

Sessions 7 and 8: Looking at Division Students are introduced to division and division notation through word problems that represent various types of division situations. They explore the relationship between the situations and the standard notations and decide how to solve problems in which the solutions do not come out evenly.

Mathematical Emphasis

- Using an array as a model for multiplication (for example, a gridded rectangle, or array, that is 3 units by 5 units models the multiplication problem 3×5)

- Becoming more familiar with multiplication pairs

- Recognizing prime numbers as those that each have only one pair of factors

- Becoming comfortable with a variety of notations used for multiplication and division (in addition to the symbols $\times$, $-$, $\div$, and $=$, students may use a multiplication symbol to represent a division situation and vice versa)

- Understanding how division notation represents a variety of division situations (including sharing situations and partitioning situations)

- Determining what to do with "leftovers" in division, depending on the situation (for example, rounding the quotient up or down or expressing the "leftovers" as a remainder, a fraction, or a decimal)

What to Plan Ahead of Time

Materials

- Interlocking cubes: at least 60 per student (Sessions 1–3)
- Chart paper (Sessions 1–3)
- Objects arranged in arrays, such as a six-pack of juice cans, carton of eggs, four-pack of yogurt cups, and card of thumbtacks (Session 1)
- Construction paper (12" by 18"): 1 per pair (Sessions 2–3)
- Scissors, glue, paste, tape (Sessions 2–3)
- Array Cards, Sets A and B, in plastic bags (Sessions 4–8). If you do not have manufactured cards, make your own; see Other Preparation.
- Calculators: at least one per pair (Sessions 7–8)
- Overhead projector, transparency pen (Sessions 1, 4–6)
- Erasable transparency pens (Session 4)

Other Preparation

- Duplicate student sheets and teaching resources (located at the end of this unit) in the following quantities. If you have Student Activity Booklets, copy only the item marked with an asterisk.

For Session 1

Student Sheet 4, Things That Come in Arrays (p. 74): 3 per pair (class), 1 per student (homework)

One-centimeter graph paper (p. 113): 1–2 per pair

For Sessions 2–3

Three-quarter-inch graph paper (p. 80): 4 per pair (class), 2 per student (homework)

Student Sheet 5, Arranging Chairs (p. 75); 1 per student (homework)

For Sessions 5–6

Game directions for the three array games (pp. 99–101): 1 copy * of each for the classroom; 1 copy of each per student (homework)

Student Sheet 6, Pairs I Know, Pairs I Don't Know (p. 76): 1 per student (homework)

For Sessions 7–8

Student Sheet 7, What Do You Do with the Extras? (p. 77): 1 per student

Student Sheet 8, What's the Story? (p. 78): 1 per student

Student Sheet 9, Word Problems (p. 79): 1 per student (homework)

- Familiarize yourself with the rules of the three array games. (Sessions 4–6)
- Make an overhead transparency of Student Sheet 4 for Session 1.
- If you have not already prepared Array Cards for use in this investigation, see What to Plan Ahead of Time for Investigation 1. (Session 4)
- Make overhead transparencies of the Array Cards for Set A, sheets 1–6 (p. 82). Then cut out each of the 51 arrays for Sessions 4–6.

Things That Come in Arrays

What Happens

Materials

- Interlocking cubes (at least 60 per student)
- Chart paper
- Student Sheet 4 (3 per pair, class; 1 per student, homework)
- Transparency of Student Sheet 4
- One-centimeter graph paper (1–2 per pair)
- Overhead projector, transparency pen
- Objects arranged in arrays, such as six-pack of juice cans, carton of eggs, pack of yogurt cups, and card of thumbtacks

Students think of items that are packaged or arranged in arrays (a rectangular arrangement of objects in rows and columns), such as juice cans or window panes. They record these types of arrays, their dimensions, and the total number of items. Their work focuses on:

- using arrays as models for multiplication
- using skip counting to count arrays
- relating multiplication notation to arrays

Ten-Minute Math: Counting Around the Class Once or twice during the next few days, do Counting Around the Class. Remember, Ten-Minute Math activities are done outside math time in any spare 10 minutes you have.

Choose a number to count by, let's say 3. Ask students to predict what number they'll land on if they count by 3's around the class exactly once. Encourage students to talk about how they could figure this out without doing the actual counting.

Then start the count: The first student says "3," the next "6," the next "9," and so forth. Encourage students to use their 100 charts if they need to. As students learn the sequence of counting by different numbers, they will rely less on their charts and more on their knowledge of the counting pattern.

Stop two or three times during the count to ask a question like this:

We're at 33. How many students have counted so far?

After counting around once, compare the actual ending number with students' predictions.

For full instructions and variations, see p. 60.

Finding Arrays in and out of Our Classroom

Groups of things often come in rectangular arrays. Take a look at this six-pack of juice cans. How many rows do you see? How many cans are in each row? Does anyone see it differently?

Students may describe this array as either three rows of two cans or two rows of three cans.

On the board or an overhead begin a chart where you can write the name of the item, write how many total items are in the array, write the dimensions of the array (written in both ways), and draw the array in both orientations. Here's a format you could use (or use a transparency of Student Sheet 4, Things That Come in Arrays):

Things That Come in Arrays

Item	Total	Dimensions		Array
six-pack of juice cans	6	2 rows of 3 3 rows of 2	2×3 3×2	
eggs	12	2 rows of 6 6 rows of 2	2×6 6×2	
yogurt containers	4	2 rows of 2	2×2	
thumbtacks	100	10 rows of 10	10×10	

❖ **Tip for the Linguistically Diverse Classroom** Show items as they are discussed—for example, pack of juice cans, eggs, yogurt containers, and thumbtacks.

What other things can you think of that come in rectangular arrays? Think about the way food is packaged at the grocery store.

What things in this room are in arrays? (window panes, cubbies, tiles, groupings of desks)

Write down some of the students' ideas on your chart. Begin to use the word *dimensions* naturally, just as you have introduced other mathematical terms into the dialogue with your class. Describe the arrays as "3 by 2" or "4 by 3," so students become used to hearing this language.

Students then work in pairs to fill in Student Sheet 4, Things That Come in Arrays (students may need several copies). Some students may need to represent the items they find or think of with cubes, tiles, or one-centimeter graph paper in order to record all the information. As you watch students work, encourage them to count their arrays using skip counting, not counting by ones.

Name _____ Date _____

Things That Come in Arrays

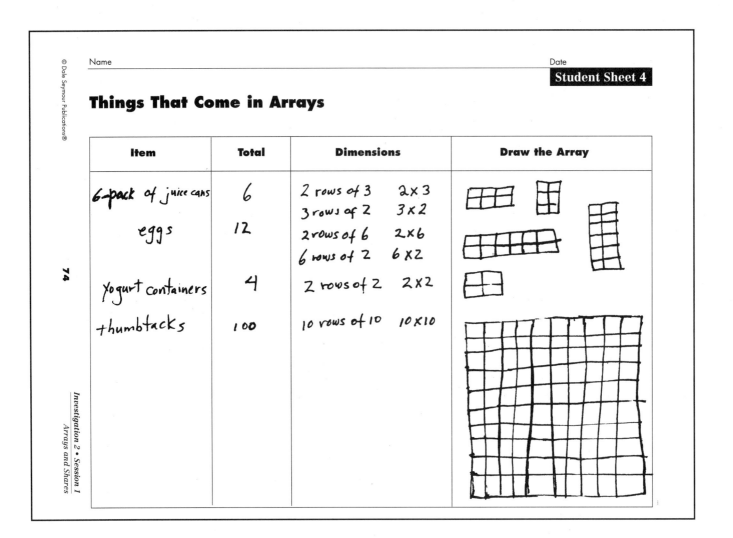

Item	Total	Dimensions	Draw the Array
6-pack of juice cans	6	2 rows of 3 2 x 3 3 rows of 2 3 x 2	
eggs	12	2 rows of 6 2 x 6 6 rows of 2 6 x 2	
yogurt containers	4	2 rows of 2 2 x 2	
thumbtacks	100	10 rows of 10 10 x 10	

Leave 10 minutes at the end of the session for students to share their findings. On chart or poster paper, begin a class list of the items students have thought of, along with the dimensions. After you have recorded several items, you might ask the following:

What items are packaged in the same types of arrays? Which arrays seem to be most popular?

What items have the same number of objects along each side? (For example, 10 × 10)

How are these arrays different from other arrays?

Throughout the rest of the unit, students can continue to search for new arrays and add them to the class list.

Things That Come in Arrays   Students look for arrays at home and record them on a copy of Student Sheet 4, Things That Come in Arrays. Some students may want to bring in arrays to show the class. As students arrive at school tomorrow, have them add to the class list new arrays they have found.

❖ **Tip for the Linguistically Diverse Classroom**   To ensure comprehension for all students, encourage everyone to bring in actual examples of the arrays listed on their charts if they can.

🏠 **Homework**

Making Arrangements

Materials

- Class array list from previous session
- Interlocking cubes (at least 60 per student)
- Three-quarter-inch graph paper (4 per pair, class; 2 per student, homework)
- 12"-by-18" construction paper (1 per pair)
- Scissors, glue, paste, tape
- Student Sheet 5 (1 per student, homework)
- Chart paper

What Happens

Using cubes and graph paper, students make all the possible arrays for a number. They make a poster of all the arrays, labeling the dimensions of each. Students discuss what they have observed about the class set of posters and explore prime and composite numbers. Their work focuses on:

- finding factors of numbers
- recognizing prime numbers as numbers with only one pair of factors
- using an array as a model for multiplication

Activity

Discussing Homework

As students arrive at school today, have them add to the class list the new arrays they found for homework. Spend some time discussing these new arrays (see the **Dialogue Box**, Things That Come in Arrays, p. 24).

What are the totals that seem to be most common? Were there any numbers smaller than 20 that you didn't find any items for? Why do you suppose that is?

Make lists of items that come in the most common amounts. List each item under the total amount.

12	3	4
eggs	tennis balls	yogurt containers
ice cubes	cupcakes	tires
pencils		

Encourage students to keep looking for arrays, especially for numbers for which no one has yet found examples.

···

❖ **Tip for the Linguistically Diverse Classroom** Instead of making a list, have limited-English-proficient students group the items brought from home by their total amounts. Then have them identify which piles are representative of items that come in the most common amounts.

···

Arranging a Group of 18 People

When a group of people need to sit down, we often arrange chairs in an array that fits the space. **Can you think of some of the places this happens?** (movie theater, bus, airplane, school auditorium)

How could we arrange chairs in a rectangular array for a group of 18 people? How else might 18 chairs be arranged in an array?

Students, working in pairs, use interlocking cubes to make all the arrays they can for 18. Ask each group to choose one of their arrays and cut it out from the graph paper.

Make sure each possible array for 18 is represented. Collect a sample of each array for 18 from the students and post them where all can see them. Some students will see 3 × 6 and 6 × 3 as two different orientations of one array; others will think of them as two arrays (see the **Teacher Note**, The Relationship Between Multiplication and Division, p. 23). Allow students to discuss their ideas about this. We suggest that you put up both arrangements, posting pairs next to each other. Arrays provide a model that helps students visualize how multiplication pairs, such as 3 × 6 and 6 × 3, are related. (If you put up all the pairs, you should have 6 items: 1 × 18, 18 × 1, 2 × 9, 9 × 2, 3 × 6, and 6 × 3.)

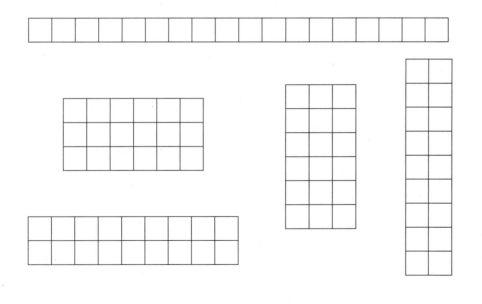

Ask students to help you label the array pairs. This is a good time to establish a class convention for which array orientation will be labeled 3 × 6 (3 rows of 6) and which will be labeled 6 × 3 (6 rows of 3). A convention will help students and you communicate with one another more clearly;

however, it is less important that students use the convention consistently than that they can clearly demonstrate what they mean.

Counting Squares in Arrays Ask students for their ideas about ways to count one of the arrays, for example, the 3 by 6 array. Some students will count the arrays by 1's, counting each individual square. Other students will see that a 3×6 array can be counted by 3's (3, 6, 9, 12, 15, 18) or by 6's (6, 12, 18). Sometimes students come up with more inventive ways, such as seeing a 3×6 as consisting of two 9's. Emphasize counting by groups by having students look at other arrays for 18. As students make arrays for other numbers in the next activity, continue to encourage them to count their arrays by groups.

Activity

Finding All Possible Arrays for a Number

During the second half of Session 2 and most of Session 3, each student works with a partner on finding ways to arrange different-size groups.

We suggest students make arrays for the following size groups: 10, 14, 15 through 30, 32, 36, 40, 42, 44, 45, 48, 49, and 50. Either assign to each pair of students one of these numbers to work on, or write the numbers on squares of paper and let students choose the number they would like. List on chart paper or the overhead the numbers that students have chosen as a way of keeping track of which numbers have been done. If they want, students can use interlocking cubes to help them plan their arrays; have all the class cubes in one place for students who may want to use them. Students should find all the possible arrays for their number and cut out each array from the three-quarter-inch graph paper. (Some students will view a 2 by 5 array, for example, as one array with two orientations. Others will still see it as two arrays—2×5, and 5×2; these students will need extra graph paper.) They glue the arrays onto a larger piece of construction paper to create a "poster" for each number with a title such as "Ways to Make 24." They should label each array with its dimensions.

Talk with groups as you observe them working:

How can you know when you have found all the arrays for one number?

Have you found any numbers that have only one array?

How can you count your array to make sure the total is correct?

Encourage students to use more than one way of counting, in order to double check the total (see the **Dialogue Box**, A Short Way to Count, p. 25). As students finish, ask them how they know they have all the arrays for their number. Encourage students to look for ways of checking their arrays by listing them in an organized way, such as 1×36, 2×18, 3×12, 4×9, 6×6. Students can post finished posters on the wall and choose another number from the list to work on. If all the numbers have been done, suggest that they choose another number (under 50) they would like to investigate.

Since students will need to see the arrays for the discussion at the end of this session, it's important to choose a place to post them where they can be viewed by all students.

Class Discussion: Looking at All the Arrays

Near the end of these two sessions, set aside about 20 minutes for students to view all the arrays, to make observations about the arrays, and to discuss, as a class, what they've noticed as they worked on their array posters. In pairs, students walk around to view all the array posters and jot down at least two discoveries they have made about the arrays in general. You might want to start their thinking by saying:

As you walk around and look at the arrays, here are some things you might pay attention to: How many arrays does each number have? Can you find out anything about even and odd numbers? See if you can make some discoveries by looking at all our arrays.

In a class discussion, students share and discuss these discoveries. Make a list of students' observations about the arrays they've constructed. Students might notice:

Some numbers have only one array which can be seen two ways, for example, 13×1 and 1×13. (Some students might say these numbers have two arrays because the two orientations look different to them.) Another way to describe these numbers is that they have only two factors, 1 and the number itself. These special numbers are called *prime numbers*. What prime numbers have students found? Do they know any above 50? Numbers having more than one array—or more than two factors—are called *composite numbers*.

Since most arrays have two factor pairs (for example, 3×4 and 4×3), most numbers have an even number of factors.

A few numbers have an odd number of factors. This occurs when one of the arrays is a square (for example, 10×10), which does not have another orientation. Some students may recognize that these numbers, such as 16, 25, and 36, are the *square numbers*.

Odd numbers never have arrays with even dimensions, but some even numbers have arrays with both even and odd dimensions. Why is this true? Why don't odd numbers sometimes have arrays with even dimensions?

Ask students: **Which of your observations do you think is always true? Which observations do you need to keep testing as you look at arrays for other numbers?**

Keep the list of their observations posted so it can be added to or revised during the course of the unit.

Sessions 2 and 3 Follow-Up

 **Homework**

Arranging Chairs Students solve more Arranging Chairs problems and write about what they notice on Student Sheet 5, Arranging Chairs, and two sheets of graph paper.

 **Extensions**

Arrays for Higher Numbers Some students may be interested in finding the arrays for higher numbers. Interesting numbers for students to investigate include 56, 60, 64, and 72.

Prime Numbers Students can also find the prime numbers from 1 to 100. This is a good activity in which to integrate the calculator.

The Relationship Between Multiplication and Division

Multiplication and division are related operations. Both involve two factors and the multiple created by multiplying those two factors. For example, here is a set of linked multiplication and division relationships:

$$8 \times 3 = 24 \qquad 3 \times 8 = 24$$
$$24 \div 8 = 3 \qquad 24 \div 3 = 8$$

Mathematics educators call all of these "multiplicative" situations because they all involve the relationship of factors and multiples. Many problem situations that your students will encounter can be described by either multiplication or division. For example:

> I bought a package of 24 treats for my dog. If I give her 3 treats every day, how many days will this package last?

The elements in this problem are: 24 treats, 3 treats per day, and a number of days to be determined. This problem could be written in standard notation as either division or multiplication:

$$24 \div 3 = \underline{} \quad \text{or} \quad 3 \times \underline{} = 24$$

Once the problem is solved, the relationships can still be expressed either as division or multiplication:

> 24 treats divided into 3 treats per day results in 8 days ($24 \div 3 = 8$)

> 3 treats per day for 8 days is equivalent to 24 treats ($3 \times 8 = 24$)

Many students in the elementary grades are more comfortable with multiplication than with division, just as they are often more comfortable with addition than with subtraction. We want students to recognize and interpret standard division and multiplication notation. However, we do not want to insist that they use one or the other to record their work when both provide good descriptions of a problem situation. In the dog-treat problem, either notation is a perfectly good description of the results.

Similarly, the order of the factors doesn't matter when describing a multiplication situation. Both of the examples that follow provide good descriptions of the dog-treat problem:

> 3 treats per day for 8 days is equivalent to 24 treats ($3 \times 8 = 24$)
> (3 per group in 8 groups is 24 total)

> 8 days with 3 treats per day is equivalent to 24 treats ($8 \times 3 = 24$)
> (8 groups with 3 per group is 24 total)

While some people prefer one or the other way to write these factors, we do not feel that a standard order (either putting the number of groups first or the number in each group first) should be taught or insisted upon. As long as students can explain their problem and their solution and can relate the notation clearly to the problem, the order of the factors in multiplication equations is not critical.

■ D I A L O G U E ■ B O X ■

Things That Come in Arrays

Students are discussing the new arrays they found for homework (p. 18).

I want to show you what Sara brought in. [*The teacher holds up an egg carton.*] **What can you tell me about this?**

Nick: It's an egg carton.

Marci: It's a 3 by 6.

Pinsuba: It can hold 18 eggs.

B.J.: You could say it has 6 rows of 3 or 3 rows of 6.

I have another egg carton over there on the table. How would you describe that one?

Kenyana: A 2 by 6.

What else can you say about it?

Kyle: Or a 6 by 2.

Irena: It holds 12.

Lesley Ann: It's like a dozen.

What do you mean by a dozen?

Lesley Ann: It's 12. A dozen is the same as 12.

Alex: My dad says a baker's dozen is 13.

How can you compare these two arrays?

Joey: Both hold eggs.

Sarah: Both count by 6's.

Kim: One's bigger.

Rebecca: One just has 6 and 6. The other one has 6 and 6 and 6.

What if I do this? (The teacher tears the egg carton for a dozen in half.)

Nick: That's a 2 by 3.

How many eggs does it hold?

Jesse: Six.

Rikki: Half a dozen.

David, want to share your album? What's it for?

David: Basketball cards. Each page is a 3 by 3.

How many cards will it hold?

David: Nine to a page, and I have almost 5 pages full.

Emilio: He'll have 45 if he gets 5 pages, because 9, 18, 27, 36, 45— that's five 9's.

Did anyone else find a different array?

Nhat: An ice tray.

Tell us about the ice tray.

Nhat: It's a 7 by 3. It has 21 holes.

That's an unusual ice tray with three rows. Any other arrays? [*Many students call out and the sharing continues.*]

⬛ D I A L O G U E ⬛ B O X ⬛

A Short Way to Count

This discussion takes place while students are working on the activity Finding All Possible Arrays for a Number (p. 20).

David: I made arrays for 24 people. One of my arrays was 6 by 4. I just thought 6 and 6 is 12, then I could see that was half of my array and just knew 12 and 12 was equal to 24.

Tyrone: You could also say 6, 12 ,18, 24. That's counting each row of 6.

Are there any other ways of counting David's array?

Shoshanna: You could count it by 4's too, but I think that takes longer than counting by 6's.

How would you count to 24 by 4's?

Tyrone: You know—4, 8, 12, 16, 20, 24.

Luisa: The 12 by 2 array is easy to count because there are 2 rows of 12, and 12 and 12 is equal to 24.

David: You could always count by 2's—2, 4, 6, 8, all the way to 24.

Shoshanna: Hey, it kind of seems like there's a short way to count and a long way. See, counting by 12's you just say 2 numbers, and counting by 2's you say a lot more. Same with 6's and 4's.

How many numbers do you have to say when you count by 2's?

[*Some students quickly count by 2's on their fingers, some look at their 100 charts, some count by 2's with cubes.*]

Preparing a Set of Arrays

Materials

- Array Cards, Sets A and B, in plastic bags (1 set per pair, class; 1 set per student, homework)
- Erasable transparency pens
- Transparencies of Array Cards, Set A
- Overhead projector

What Happens

Students prepare a personal set of Array Cards for use in homework assignments throughout the remainder of the unit. They label the Array Cards in Set A, consisting of 51 arrays, which represent the multiplication combinations of the factors 2 through 12 with totals (or products) up to 50. Later in the unit, or as they become fluent working with these cards, students gradually add the larger arrays in Set B. Students begin to learn several games that use the Array Cards. Their work focuses on:

- using an array as a model for multiplication
- becoming more familiar with multiplication pairs

Activity

Preparing a Personal Set of Array Cards

Tell students they are going to have a class set and an at-home set of Array Cards for multiplication pairs. They will use the cards to play games and to practice multiplication problems. During the first few days, students will use arrays for totals up to 50 (Array Cards, Set A). As they become more comfortable with these pairs, they can add arrays for totals up to 144 (Array Cards, Set B).

If you have enough manufactured Array Cards for student pairs to use in class and individual students to use at home, classroom aides or parent volunteers will have already labeled the class sets of cards in *erasable* transparency pen, as described below. Students can do the same to their at-home sets as a classroom activity.

Other senarios: If you have enough manufactured cards for only the class sets, aides or parent volunteers will have already labeled them as described below and will have used the blackline masters in the back of this book to make at-home Array Cards for the students to fully label as described below. (How to Make Array Cards on p. 81 tells how to use blackline masters to make Array Cards and how to label them.) If you have no manufactured Array Cards, aides or volunteers will have already used blackline masters to make and fully label a class set per pair of students and to make at-home sets for individual students themselves to fully label. In case the sets of Array Cards that go home with students are the only ones you have, be sure that students return the cards each day for classroom use.

Labeling Array Cards For manufactured Array Cards, aides or students use *erasable* transparency pens to label the nongrid side of each card with one dimension of the array, as a hint to help students when the arrays are new. Students can erase this hint when they feel more confident. For cards made from blackline masters, aides or students use pencils, not markers or crayons, so the numbers cannot be seen through the paper. They label the grid side of the card with the dimensions of the grid (for example, 4 × 6, 6 × 4). On the other side of the card (the blank side), they label the total number of squares in the grid. Again, they may also label one dimension lightly as a hint to help students when the arrays are new. Each set of labeled cards, manufactured or not, is stored in a plastic bag with the name of the student pair or individual student on it.

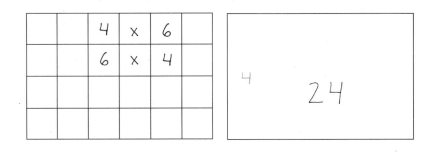

Using the Array Cards

Students use their array cards to play two multiplication and division games: Multiplication Pairs and Count and Compare. See directions for these games on pp. 99 and 100. The **Teacher Note**, Array Games, on p. 28, highlights the kinds of skills emphasized by these games.

The games can be taught to the whole class or to small groups of students. To begin, you might teach the Multiplication Pairs game to pairs of students after they complete their array cards. These students can then teach other students.

Ten minutes before the end of this session have the students stop working so that you can introduce the Count and Compare game to the whole class. You can play this game on the overhead or by taping array cards to the board. As you demonstrate the game, ask students for ideas about how the arrays can be counted. Encourage students to count by groups, looking at the number of squares in a row or column. Ask students if they can determine which array has a bigger total by comparing the dimensions of two arrays.

Suppose you were comparing two arrays and did not know the total of either array. How could looking at the dimensions of the arrays help you decide which one was bigger? What if the length of one array is bigger than the other but the width is less? How might you compare the two to determine which array had more squares?

Most students need to place the arrays next to each other or on top of each other to match the sides and count the squares that overhang.

If there is time, let the students play a few rounds of Count and Compare in pairs. During the next two sessions and in the next investigation students will have time to play these array games and learn others.

Teacher Note ▷ *Array Games*

Students use their sets of arrays to play multiplication and division games over the course of many days. Using the arrays in this way gives students a variety of opportunities to use a concrete model for multiplication and division, and in the process of playing the games, they become more familiar with the "multiplication pairs."

The Multiplication Pairs game focuses on practicing the problems by representing them as rectangular arrays. By looking at the dimensions (for example, 8×6), students can skip count the rows or columns or do repeated addition to figure the total amount. When they look at the total (48) they need to think about the possible multiplication combinations (4×12 or 6×8) that make 48, and select the combination that best matches the dimensions of the rectangle. They are often thinking of division here: "I know I can skip count by 4's to 48; let's see, how many 4's is it?" By recording the pairs they know and don't know, students keep track of their own progress and can be encouraged to set reasonable goals for learning new multiplication pairs.

The game Count and Compare provides students with opportunities to compute in a variety of ways the total number of squares in an array. Some students may choose to skip count the rows in order to figure the total. Some students will know just certain pairs and not need to figure the total. Other students might look at the

dimensions of the two arrays, comparing the length of rows and columns. Because there are various ways to count and compare the arrays, this game allows students with varying experience with multiplication and division pairs to play together. You might need to establish some guidelines—for example, each person has to compute his or her own total—as a way of allowing each student the time and opportunity to think about each array.

The Small Array/Big Array game encourages students to build a larger array from two or three smaller arrays. By doing this, they create a concrete model for how a multiplication situation such as 7×9 can be pulled apart into smaller, more manageable components, such as 3×9 and 4×9. As they record each match in the form of an equation, they connect the concrete model with the equations that represent these relationships.

All students should have the opportunity to play each game more than once during the course of this unit. Having students play array games in pairs keeps the pace of the game moving. As students play the games and use their arrays, they become more familiar with the multiplication and division pairs each array represents. They play the games during Investigations 2 and 3 and should take their paper set of arrays home to play with family members for homework.

Array Games

What Happens

During these two sessions students are introduced to Choice Time. They have the opportunity to play three array games and to practice skip counting in pairs or small groups. As they become more comfortable and familiar with these arrays, they add larger numbers to their sets until the sets represent all multiplication combinations with factors 2 through 12. Their work focuses on:

■ becoming more familiar with multiplication and division pairs through arrays and skip counting

Ten-Minute Math: Counting Around the Class Once or twice during the next few days, continue to do Counting Around the Class.

Count by 4's or 5's. Before you begin, ask:

If we count by 4's (or 5's) around the class, do you think that the number we end on will be more or less than 100? Why do you think that?

Stop two or three times during the count and ask questions like these:

We're at 65 now. How many students have counted?

How many more students will have to have a turn before we reach 100?

On some days you might have everyone use a calculator or have a few students use calculators to skip count while you are counting around the class.

For full instructions and variations, see p. 60.

Materials

■ Array Cards, Sets A and B, in plastic bags (1 set per pair, class; 1 set per student, homework)

■ Highlighted 100 charts from Investigation 1

■ Overhead projector, transparency pen

■ Transparencies of Array Cards from previous session

■ Student Sheet 6 (1 per student, homework)

■ How to Play Multiplication Pairs (1 per student, homework)

■ How to Play Small Array/Big Array (1 per student, homework)

■ How to Play Count and Compare (1 per student, homework)

Activity

Another Array Game

Begin today's session by introducing the Small Array/Big Array game. (Directions for this game can be found on p. 101.) You can introduce this game to the entire class by playing with a small group of students, or you can teach the game to one group of students at a time while the others play one of the array games introduced in the last session. Because this game is played on a flat surface, you might want to gather the students in a circle on the floor or around a group of desks to demonstrate.

Display all the arrays. Ask students to find two arrays that could be used to "match" a bigger array. For example:

If you are trying to cover a 5 × 7 array and have a 2 × 7 array, what array would you need to complete the match?

As "matches" are completed, ask students to express the match using mathematical notation and check it by comparing the totals of the smaller arrays with the bigger array. For example, in the above situation, students would write: 5 × 7 = 2 × 7 + 3 × 7; this checks because 35 = 14 + 21.

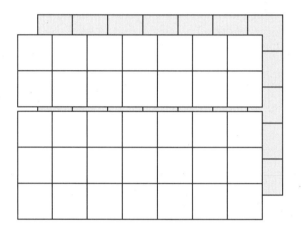

Once students grasp the idea of this game, allow them to play with partners.

Choice Time: Multiplying, Dividing, and Skip Counting

Four Choices For the remainder of this session and the next, students choose from four multiplication and division activities going on simultaneously in the classroom. This is referred to as Choice Time. This format allows students to explore the same idea at different paces. Students may repeat an activity, using different numbers or problems. You might want to establish an expectation with your students that they play at least two array games and practice skip counting with their partners by the end of tomorrow's session. Many students will have extra time and may choose to play all of the array games or just one game many times.

These activities are designed to give students more experiences using models, such as the 100 chart, and arrays to help them develop their knowledge of the multiplication and division relationships.

How to Set Up the Choices Write the four choices on the board with a list of the materials needed for each choice, as a reminder for the students.

If you set up your choices at stations, list the materials students will find at each station. Students can keep track of their choices in their own way.

Choice 1: Array Game: How to Play Multiplication Pairs; Array Cards, Sets A and B (1 set per pair); paper and pencil

Choice 2: Array Game: How to Play Count and Compare; Array Cards, Sets A and B (1 per pair)

Choice 3: Array Game: How to Play Small Array/Big Array; Array Cards, Sets A and B (1 set per pair); paper and pencil

Choice 4: Skip Counting: students' highlighted 100 charts from Investigation 1

Introduce Choice 4, Skip Counting, which is described below. Choices 1, 2, and 3 are array games with which the students are already familiar. Post a copy of each set of game rules so that students may refer to the directions when in doubt about the rules of the game. Leave the rules posted for use in later class sessions. Students may choose to play using only the Array Cards in Set A, which consists of multiplication pairs with products up to 50. Then when they feel comfortable, students may include Array Cards from Set B.

Choice 1: Array Game: Multiplication Pairs

Given the dimensions of an array, students are to find the total number of squares in the array; given the total, students are to find the dimensions. As they play, students write the multiplication pairs and relationships they know and don't know on a sheet of paper.

Choice 2: Array Game: Count and Compare

Students use multiplication relationships to find the sizes of students' array cards and then determine the largest.

Choice 3: Array Game: Small Array/Big Array

Students use their array cards to make "matches" between a large array and two or more smaller arrays. The players should write their "matches" on a sheet of paper using mathematical statements.

Choice 4: Skip Counting

With a partner, students practice skip counting by 2's through 12's, using the highlighted 100 charts they made in Investigation 1, if they need to. One way to do this is for one student to look at a chart while the other student counts without seeing the chart; the first student checks the count and provides hints as necessary. You might suggest that students who choose this activity skip count by 2's, 3's, and 6's and keep a record of these and other numbers they count by.

Teacher Checkpoint

Choice Time

During Choice Time, circulate among the groups and observe students as they are involved with an activity, or use the time to meet with small groups of students who are having difficulty with a particular activity.

Some things you might look for are the following:

- How are students making decisions about choosing an activity and organizing their time and materials?
- Are there too many or not enough activities going on at once?
- Are students keeping track of the choices they have completed?
- How are students figuring out the total number of squares in arrays? Are they counting one by one? Counting by groups? Do they know the multiplication pairs?
- Do some students need to spend more time counting by 2's, 3's, and 6's?
- Are some students ready to add the next set of arrays (Set B) to their existing set?

Students will continue working on all the activities throughout the next investigation.

Some students may need help in making choices and keeping track of what they need to do and what they have completed during Choice Time. It will probably take a few sessions before students are feeling more confident with the routines and expectations. You may ask for student volunteers to act as resources for classmates who need help or clarification.

You can make adjustments in the number of choices offered, the number of array cards that students work with, and the overall pacing of Choice Time in order to best meet the needs of your classroom.

Activity

How Many Ways Can You Make 8 × 6?

At the end of Session 6, bring the class together for a discussion about the Small Array/Big Array game. Students will need their sets of array cards.

As you played Small Array/Big Array, what clues helped you decide which two smaller arrays might match up to make a bigger array?

Let's take a look at the 8 × 6 array. What are some of the combinations of smaller arrays that would equal this array? Use your set of arrays to make as many 8 × 6 arrays as you can using either two or three arrays.

Every student pair builds 8×6 arrays on their desks. When most students
have constructed a few examples, ask students to share their combinations
of arrays by describing the dimensions of the smaller arrays. Record
these on the board or overhead as shown below. If you have made a set
of overhead arrays, you can display the examples as well as record the
dimensions.

Challenge students to find all the ways to make 8×6 using two arrays. If
you record combinations in an ordered list such as the one below, students
may more easily use the patterns to exhaust all possible combinations.

$8 \times 6 = 7 \times 6 + 1 \times 6$ $8 \times 6 = 8 \times 1 + 8 \times 5$
    $48 = 42 + 6$ $48 = 8 + 40$

$8 \times 6 = 6 \times 6 + 2 \times 6$ $8 \times 6 = 8 \times 2 + 8 \times 4$
    $48 = 36 + 12$ $48 = 16 + 32$

$8 \times 6 = 5 \times 6 + 3 \times 6$ $8 \times 6 = 8 \times 3 + 8 \times 3$
    $48 = 30 + 18$ $48 = 24 + 24$

$8 \times 6 = 4 \times 6 + 4 \times 6$
    $48 = 24 + 24$

Here are the ways to construct 8×6 using two arrays:

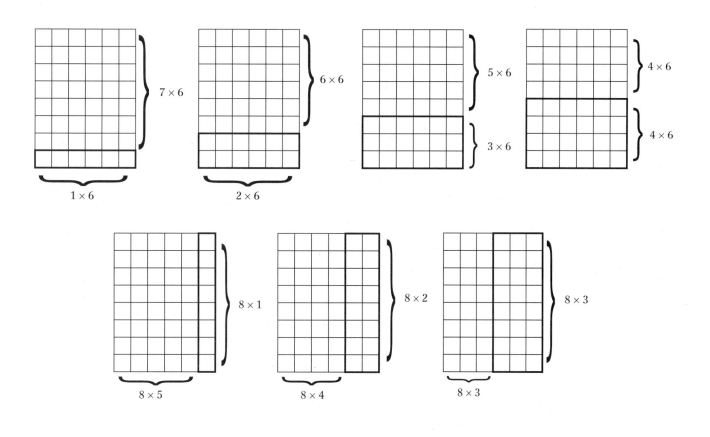

Emphasize the idea that unfamiliar multiplication situations can be broken down into smaller, more manageable components:

If you don't know how much 12 × 4 is equal to, how could you break it down into more familiar parts that you know?

Sessions 5 and 6 Follow-Up

Homework

Pairs I Know and Pairs I Don't Know   After Session 5, students play Multiplication Pairs at home and record "Pairs I Know" and "Pairs I Don't Know" on Student Sheet 6. Students will also need the directions, How to Play Multiplication Pairs (p. 99), and their personal set of Array Cards.

Note: If students' paper sets are the only Array Cards you have, students must remember to bring them back to school each day during the course of this unit.

Array Games   After Session 6, students play one of the Array Games with someone at home. Send home the directions for Count and Compare and Small Array/Big Array (pp. 100–101). Suggest that students keep the directions in a safe place for future reference.

Looking at Division

What Happens

Students are introduced to division and division notation through word problems that represent various types of division situations. They explore the relationship between the situations and the standard notations and decide how to solve problems in which the solutions do not come out evenly. Their work focuses on:

- recognizing division situations
- using division notation
- solving division problems
- expressing remainders using problem context

Materials

- Array Cards, Sets A and B (1 set per pair)
- Calculators (at least 1 per pair)
- Student Sheet 7 (1 per student)
- Student Sheet 8 (1 per student)
- Student Sheet 9 (1 per student, homework)

Activity

Ask students to get out all the arrays that have 36 total squares. Write on the board:

$$36 \div 4$$

Here's a division problem. How do you read this? Which array would help you solve it?

Creating Problems   Students create their own problem situations to model the above division expression.

Can you think of a problem situation that you'd write down as 36 ÷ 4?

Students may think of division problems that involve sharing ("There are 36 marbles being shared among 4 friends. How many marbles will each person get?") and division problems that involve grouping or measuring ("There are 36 marbles. I'm going to put 4 marbles in each bag. How many bags will I need?"). See the **Teacher Note**, Two Kinds of Division: Sharing and Partitioning (p. 39) for more information about these two division situations.

Introducing Division Notation

Introduce the notation: 4)‾36‾

Discuss with students how this is read, and point out the total that is being divided appears as the second number, whereas when they use the ÷ notation, it is the first number. (See **Teacher Note**, Talking and Writing About Division, p. 41.)

How would you use your calculator to solve this problem? Which keys would you use? Which number would you put in first?

Students try to solve the problem on their calculators and share their strategies and solutions.

Creating More Problems Ask students to look at their arrays to find other ways to divide 36.

How else can you divide up 36? Who can think of a new problem situation that involves dividing up 36?

Students spend a few minutes working in pairs to generate division problems that start with the total 36. Have them:

1. Write down each problem situation (for example, "There are 36 children who are going to divide up into 6 teams for relay races. How many will be on each team?").

2. Write down the notation for that problem using both ÷ and)‾‾ .

3. Give the solution to their problem for both the notations ÷ and)‾‾ .

❖ **Tip for the Linguistically Diverse Classroom** Pair English-proficient students with limited-English-proficient students. After partners have conferred with each other about possible ideas, have English-proficient students write the problem. Then have both partners write the notation and solve the problem.

As you circulate, observe whether students are comfortable creating division problems and whether they understand the correspondence between their problem situation and the written notation. Help students to read notation in context; for example, the expression $36 \div 9$ may be read as "36 divided into 9 groups" in a sharing situation or as "how many 9's are in 36?" in a partitioning situation.

Ask some students to present their problems. Other students can demonstrate how to express the problems using standard notation and can give the solutions.

Activity

Division Problems with Leftovers

Write on the chalkboard:

$36 \div 5$

Look at this problem. What would be a situation in which you would have 36 divided by 5?

List students' answers on the board. Then using students' situations to illustrate this problem, discuss how you would find the solution to $36 \div 5$ and what would happen to the "extra." Students may express this extra as a fraction, a decimal, or a remainder or leftover amount.

Have students work in groups on Student Sheet 7, What Do You Do with the Extras? Tell students that these are all division problems where you can't divide the total evenly. Their job is to decide what to do in each case and to write down their reasoning and a reasonable solution. Students can use their arrays and calculators.

❖ **Tip for the Linguistically Diverse Classroom** Form groups that represent a mixture of limited-English-proficient students with English-proficient students to work on Student Sheet 7.

Class Discussion: Division with "Extras"

Ask students to share their solutions to the division problems on Student Sheet 7. Invite some students to write their solutions with explanations on the board. Sample solutions may include:

There are 36 people to be transported in vans that hold 8 people. How many vans do you need?

> There are 4 vans with 4 people left.
> So you'd need 5 vans to take all the people.

There are 36 crackers that will be shared by 8 people. How many crackers will each person get?

> Each person can get 4 crackers. Keep 4 crackers for another day.
> $36 \div 8 = 4$ crackers per person with 4 extras
> > or
> Each person can get $4\frac{1}{2}$ crackers. $36 \div 8 = 4\frac{1}{2}$

There are 36 people in rows of 8. How many rows will you need?

> There are 32 people who will fit in 4 rows. Then 4 people have to sit in row 5. $36 \div 8 = 4$ rows, with 4 more people
> > or
> You fill up 4 rows and half of another row. $36 \div 8 = 4.5$

Assessment: Creating and Solving Division Problems

Students work individually on Student Sheet 8, What's the Story? Explain that students are to think of a division situation for each of the expressions on the student sheet. As you observe students working and look at their finished papers, consider the following:

1. Does this student understand what kinds of situations are represented by written division notation? Does this student include both sharing and grouping situations in the problems?
2. Can this student solve division problems accurately? Does the student use the array cards or the calculator appropriately to solve these problems? Can the student do many of these division problems mentally?
3. Can this student give a reasonable solution when the total cannot be divided evenly into equal groups?
4. Is this student able to use fractions or decimals appropriately in the solution to Problem 2 (and Problem 4)?

❖ **Tip for the Linguistically Diverse Classroom** Have limited-English-proficient students use pictures or drawings to convey their story problems.

Word Problems   After Session 8, students take home Student Sheet 9, Word Problems, to complete as homework.

Two Kinds of Division: Sharing and Partitioning

> **Teacher Note**

There are two distinct kinds of division situations. Consider these two problems:

> I have 18 balloons for my party. After the party is over, I'm going to divide them evenly between my sister and me. How many balloons will each of us get?

> I have 18 balloons for my party. I'm going to tie them together in bunches of 2 to give to my friends. How many bunches can I make?

Each of these problems is a division situation—a quantity is broken up into equal groups. The problem and the solution for each situation can be written in standard notation as $18 \div 2 = 9$.

Although similar, these two situations are actually quite different. In the first situation, you know the number of groups—2. Your question is, "How many balloons will be in each group?" In the second situation, you know that you want 2 balloons in each group, and your question is, "How many groups will there be?" In each case you divide the balloons into equal groups, but the results of your actions look different.

I have 18 balloons and 2 people. How many for each person?

I have 18 balloons to put in bunches of 2. How many bunches?

Continued on next page

The first situation is probably the one with which your students are most comfortable, because it can be solved by dealing out. That is, the action to solve the problem might be: one for me, one for you, one for me, one for you, until all the balloons are given out. In this situation, division is used to describe sharing.

In the second situation, the action to solve the problem is making groups—that is, taking out a group of 2, then another group of 2, and another, and so on, until no balloons are left. In this situation, division is used to describe *partitioning*.

Students need to recognize both of these actions as division situations and to develop an understanding that both can be written in the same way: 18 ÷ 2 = 9. Therefore, in this unit we present both kinds of division problems and, depending on the situation, help students interpret the notation as either "How many 2's are in 18?" or "Divide 18 into 2 groups; how many in each group?"

As students become more flexible with division, they will understand that they can think of a sharing situation as partitioning, or a partitioning situation as sharing, in order to make it easier to solve. For example:

> How many will be in each team if I make 25 teams from 100 people?

To solve this, I will probably not deal out 100 into 25 groups. It is easier to think, "How many groups of 25 are in 100?"—knowing that the answer to this partitioning question will also give me the answer to the sharing problem I need to solve. Some of your students may soon have an intuitive understanding that they can divide in either way to solve a division problem.

Also, as students get more experience with both multiplication and division, they will begin to recognize how related problems can help them. Just as $2 \times 9 = 9 \times 2$, $18 \div 2 = 9$ is already related to $18 \div 9 = 2$. When you find students stuck in routines that are difficult to keep track of—such as dealing out 100 into 25 teams—ask if thinking about a similar problem might help them. "If there were 100 people in all and 25 people on a team, how many teams would there be? Does that give you any ideas about making 25 teams from 100 people?"

Talking and Writing About Division

Various division symbols are used as standard notation in our society:

$$24 \div 4 \qquad 4\overline{)24}$$
$$24/4 \qquad \frac{24}{4}$$

We want students to recognize these forms (which they may see on tests and in other textbooks) as having the same meaning. They will use the fractional form in the Fractions units of the *Investigations* curriculum.

There are many different ways to "read" or speak of these notations:

Four goes into (or, as students say, "guzinta") 24

24 divided by 4

How many 4's are in 24?

When 24 is shared among 4 people, how many does each person get?

So many symbols and so many different ways of reading them can be very confusing to young students, especially because the numbers and symbols appear in different positions, depending on which notation you are using. We would like students to read division notation with as much meaning as possible, so that they connect the symbols to the situations they represent.

The first two ways (above) of "reading" these notations correspond to the ways the symbols are written, and may seem simpler to you. However, "Four goes into 24" is a phrase that carries little meaning about the division operation. Encourage students to interpret these symbols as "24 divided into 4 parts, or "How many 4's are in 24?" These two phrases express more meaningfully what the notation represents.

Talk explicitly with your students about what these symbols mean, and how it can be confusing to read them. Work with them to help them remember that the quantity being divided is always the first number in the form $24 \div 4$, and always the inner number in the form $4\overline{)24}$. Don't let them rely on thinking of the bigger number

as the one being divided, since $1 \div 2$ (for example) is a perfectly legitimate division problem, in which a smaller number is divided by a larger number.

Sometimes a problem does not divide evenly. Rather than teaching students to write "R" for the remainder, have them describe the remainder in any way that makes sense to them for that problem. For example, how many groups of 3 can be formed with 26 students? Some students may decide they can make 8 groups of 3 and one group of 2. Others may decide to make 6 groups of 3 and two groups of 4. If the example was 26 cookies to share among 3 children, they might give 8 to each and leave the remaining 2 cookies on the plate, or break them up to share.

For more about helping students connect their own good strategies with standard notation, see the **Teacher Note**, What About Notation? (p. 11).

Multiplication and Division with Two-Digit Numbers

What Happens

Session 1: Multiplication Clusters Students solve groups ("clusters") of related multiplication problems. They look for patterns and relationships within each cluster that can help them solve the last problem and then write about their problem-solving strategies.

Sessions 2, 3, and 4: Multiplication and Division Choices Students choose from a selection of multiplication and division activities, including array games, skip counting, and solving multiplication clusters, and division word problems.

Session 5: Problems That Look Hard But Aren't In the final session of this unit, students work in small groups to generate a list of "hard" multiplication pairs (or relationships). The class compiles a list of these pairs and discusses which

are the most difficult to learn and which they have developed strategies for solving. They solve a two-digit multiplication problem as a whole class. As an assessment activity, students solve another two-digit problem in two different ways.

Mathematical Emphasis

- Becoming fluent in basic multiplication relationships
- Partitioning numbers to multiply them more easily (for example, 7×23 can be 7×10 plus 7×10 plus 7×3)
- Recognizing multiplication and division situations and representing each situation using a mathematical statement
- Learning about patterns that are useful for multiplying by multiples of 10 (multiplying by 10, 20, . . . , 100)

What to Plan Ahead of Time

Materials

- Interlocking cubes: at least 60 per student (Sessions 1–5)
- Array Cards, Sets A and B: 1 per pair (Sessions 2–4)
- Overhead projector, transparency pen (Sessions 2–5)

Other Preparation

- Duplicate student sheets and teaching resources (located at the end of this unit) in the following quantities. If you have Student Activity Booklets, no copying is needed.

For Session 1

One-centimeter graph paper (p. 113): 1–2 per pair

Student Sheet 10, Another Set of Related Problems (p. 102): 1 per student (homework)

For Sessions 2–4

Student Sheet 11, Multiplication Clusters (Sets A–J), (p. 103): 1 complete set per student

Student Sheet 12, Problems About Our Class (p. 108): 1 per student

Student Sheet 13, Recycling Problems (p. 109): 1 per student

Student Sheet 14, Multiplication Clusters at Home (p. 110): 1 per student (homework)

For Session 5

One-centimeter graph paper (p. 114): 1–2 per student

Student Sheet 15, Two Ways to Solve a Problem (p. 111): 1 per student

Student Sheet 16, Problems That Look Hard But Aren't (p. 112): 1 per student (homework)

- Be sure directions are still posted for array games: Small Array/Big Array, Multiplication Pairs, and Count and Compare (pp. 99–101). (Sessions 2–4)

Multiplication Clusters

Materials

- Interlocking cubes (at least 60 per student)
- One-centimeter graph paper (1–2 per pair)
- Student Sheet 10 (1 per student, homework)

What Happens

Students solve groups ("clusters") of related multiplication problems. They look for patterns and relationships within each cluster that can help them solve the last problem and then write about their problem-solving strategies. Their work focuses on:

- using multiplication relationships
- breaking down large problems

Activity

Solving Cluster Problems

Remind students what cluster problems are (you may want to read the **Teacher Note** Cluster Problems on p. 47).

For the next few days we'll be thinking about good strategies for figuring out hard multiplication problems. We'll be looking at groups of multiplication problems that are related to one another. These groups of problems are called Cluster Problems.

When we did Cluster Problems before, you used the first two problems to help you solve the third problem. These clusters sometimes have four or five problems in each cluster. Again, you'll use the earlier problems to help you solve the last problem in the cluster. You can use the interlocking cubes or the graph paper to make arrays to help you solve these problems. But try to solve the last problem by thinking about the other problems in the cluster. You can add problems to the cluster that help you solve the final problem better.

Present the following cluster:

$$4 \times 5$$
$$2 \times 15$$
$$4 \times 10$$
$$4 \times 15$$

Students talk in small groups about how to solve this first cluster. They can work on the first three problems in any order, then decide how to use one or more of those problems to solve 4×15. When they have finished, have volunteers show how they did the problems to the whole class.

Present one more cluster to the class.

```
2 × 8
6 × 8
10 × 8
12 × 8
```

Students complete the problems and compare strategies within their groups. As the students are working, circulate among the groups and note the strategies students are using.

Sharing Strategies   Bring the whole class together briefly to share strategies they have figured out for themselves or learned from one another. Possible questions you might ask to help focus their thinking are:

Did you learn something useful from someone else in your group? Tell us about it.
Which problems did you use to solve the last problem?
Did anyone solve the problem in a different way?
Did anyone add a different problem to the cluster to help solve 12 × 8?

Writing About Your Strategies

Show a third set of problems. Suggest to students that they solve mentally any of the first three problems they can and write down the answers, then use those answers to solve the rest of the problems, including the last problem in the cluster. As always, they should write down any different problems they use to help them solve 3 × 24.

```
3 × 10
3 × 20
3 × 4
3 × 24
```

This time, students work alone to solve the problems. They write about how they solved them, then create a set of arrays to show their thinking.

❖ **Tip for the Linguistically Diverse Classroom**   Have limited-English-proficient students demonstrate their thought processes by illustrating their strategies. Omit instructions that direct students to write about their approach.

One way to solve 3×24 is with centimeter graph paper:

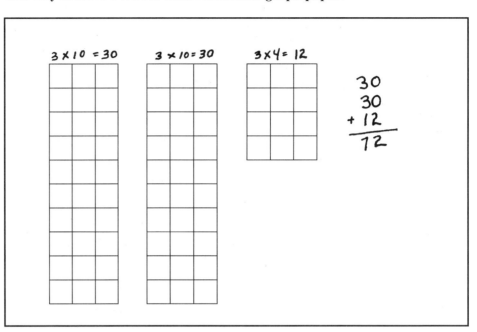

The **Dialogue Box,** Ways to Solve 3×24 (p. 48), gives some examples of student strategies for solving this multiplication cluster.

Session 1 Follow-Up

Another Set of Related Problems   Assign the set of cluster problems on Student Sheet 10, Another Set of Related Problems.

Students find the solution to each problem and write about how they solved the last problem. Provide time in the next session for students to compare their solutions and strategies with classmates. Encourage students to invent their own set of cluster problems they can share with classmates.

❖ **Tip for the Linguistically Diverse Classroom**   Have limited-English-proficient students demonstrate their solutions by illustrating their strategies as shown in the example on this page. Omit the instructions that direct them to write about their approach and to invent their own set of cluster problems.

Cluster Problems

Cluster problems are sets of problems that help students think about using what they know to solve harder problems. For example, what do you know that would help you solve 12×3? If you know that $3 \times 3 = 9$, you might double that solution to get the solution to 6×3. Now that you know that 6×3 is 18, you can double that to get 12×3. Or, you might start with 10×3. If you know that $10 \times 3 = 30$, then you can start with 30 and add two more 3's to get 36. As students work with clusters, they learn to think about all the number relationships they know that might help them solve the problem they are working on.

The cluster problems used in this unit are designed to help students make sense of multiplying two-digit numbers. They build an understanding of the process by pulling apart multiplication problems into manageable subproblems,

solving each of the smaller problems, then putting the parts back together. This process is based on an important characteristic of multiplication called the *distributive property.* You may have learned the name of this property in your own schooling without ever quite understanding what it is. In this unit, we don't teach students the name of the property, but it *is* a core idea of the unit. Here is an example: 9×23 can be thought of as $(9 \times 10) + (9 \times 10) + (9 \times 3)$. In this example, 23 is pulled apart into $10 + 10 + 3$, and *each part* must be multiplied by 9 in order to give the solution to 9×23. The number that we split up doesn't have to be split up into 10's and 1's. For instance, 8×12 could be taken apart into $(4 \times 12) + (4 \times 12)$ or into $(8 \times 6) + (8 \times 6)$. In each case, one of the numbers is split up into parts, and each part must be multiplied by the other number in order to maintain equivalence to the original expression–$8 \times 12 = (4 \times 12) + (4 \times 12)$ or $8 \times 12 = (8 \times 6) + (8 \times 6)$.

As students solve the first few problems in each cluster, they continue to become familiar with the single-digit multiplication pairs. Students will begin to say "I just knew it" for some of these pairs, as they become part of their known repertoire of multiplication combinations. They will also make use of multiplying by 10 and by multiples of 10, another essential tool in solving harder multiplication problems (see the **Teacher Note**, Multiplying by Multiples of 10, p. 54).

Cluster problems are intended to help students learn how to look at a problem and build a strategy to solve it based on the number relationships they know. At first, students work on clusters of problems that are provided for them. Later in this unit, students begin to create their own clusters of problems. (In later units of *Investigations* in both grades 4 and 5, students spend more time creating their own clusters of problems as well as using a variety of given problems to solve harder computation.) Throughout their work on cluster problems in this unit, encourage students to add to the clusters any problems that they use to solve the final problem in the cluster.

SET C

$2 \times 5 = 10$

$3 \times 5 = 15$

$10 \times 5 = 50$

$30 \times 5 = 150$

$32 \times 5 = 160$

on the last one I knew that $30 \times 5 = 150$ so I did

$5 \times 2 = 10$
$30 \times 5 = 150$
$150 + 10 = 160$

SET D

$5 \times 7 = 35$

$10 \times 7 = 70$

$4 \times 25 = 100$

$20 \times 7 = 140$

$25 \times 7 = 175$

It was easier to use 25's, so I used $4 \times 25 = 100$, then you just need 3 more 25's, 125, 150, 175.

Ways to Solve 3 × 24

The class has been working on the following cluster problem in the activity Writing About Your Strategies (p. 45). They have written about how they have solved the cluster and are sharing their strategies for thinking about 3 × 24 in a whole-class discussion.

$$3 \times 10$$
$$3 \times 20$$
$$3 \times 4$$
$$3 \times 24$$

Kumiko: The way I did these was to count by each number. For 3 × 10, I counted by 10's. For 3 × 20, I counted by 20's. I just knew 3 × 4 was equal to 12. Then for 3 × 24, I added three 24's.

Rafael: I knew that 3 × 20 was 60 from the second problem. I knew that 3 × 4 was equal to 12, then I added the two together and got 72.

Ahmad: The way I figured out 3 × 24 was I know 24 plus 24 is equal to 48, so I added 2 from the next 24 and that made 50. Then I had to add the 22 that was left to the 50, and this gave me 72.

Shoshana: At first I didn't know how to do 3 × 24, but I knew that 3 × 20 = 60. So I tried to add 4 to the 20 and that made 24, *[Shoshana wrote 3 × 24 = 60]*. Then I added 4 to the 60 and got 3 × 24 = 64. But now I can see I made a mistake because I should have added three 4's to 60, not just one.

Vanessa: Mine was sort of like Ahmad's way. I just knew that 24 plus 24 is equal to 48, then I added 24 more and got 72.

Joey: I counted by 24's: 24, 48, 72!

Nick: I took 60 from 3 × 20, then I added 4 more 3's and got 72.

Multiplication and Division Choices

What Happens

Students choose from a selection of multiplication and division activities, including array games, skip counting, and solving multiplication clusters and division word problems. Their work focuses on:

- recognizing multiplication and division situations
- using multiplication and division notation
- solving multiplication and division problems

Ten-Minute Math: Multiple BINGO During the next few days, have students spend time outside of math class playing Multiple BINGO.

Using 100 charts, students try to color in 5 numbers in a row.

A factor card is turned over from the deck, and each player may color in any multiple of that factor on the 100 chart.

If a player turns over a Wild Card, that player decides what factor it represents. For a full description of Multiple BINGO, see the game directions on p. 62.

Materials

- Array Cards, Sets A and B (1 set per pair)
- Interlocking cubes (class set available)
- Student Sheet 11, Sets A–J (1 complete set per student)
- Student Sheet 12 (1 per student)
- Student Sheet 13 (1 per student)
- Game directions (already posted)
- Overhead projector, transparency pen
- Highlighted 100 charts from Investigation 1
- Student Sheet 14 (1 per student, homework)

Activity

Six Choices During Sessions 2, 3, and 4 of this investigation, students may choose from the six multiplication and division activities listed on p. 50. Students may have worked on the first four of these activities in Choice Time: Multiplying, Dividing, and Skip Counting (p. 30). If you and your students have not done a Choice Time before, it will be important to establish some guidelines during this session.

How to Set Up the Choices Introduce the six activities and write the six choices on the board as a reminder for the students. If you set up your choices at stations, show students what materials they will find at each station. Students can keep track of their choices in their own way.

Choice Time: Multiplication and Division Activities

Choice 1: Skip Counting: students' highlighted 100 charts from Investigation 1

Choice 2: Array Game: How to Play Small Array/Big Array (already posted); Array Cards, Sets A and B (1 set per pair); paper and pencil

Choice 3: Array Game: How to Play Count and Compare (already posted); Array Cards, Sets A and B (1 set per pair)

Choice 4: Array Game: How to Play Multiplication Pairs (already posted); Array Cards, Sets A and B (1 set per pair); paper and pencil

Choice 5: Multiplication Clusters: Student Sheet 11, Multiplication Clusters, Sets A–J (1 complete set per student); pencil

Choice 6: Solving Division Word Problems: Student Sheet 12, Problems About Our Class (1 per student); Student Sheet 13, Recycling Problems (1 per student); pencil

Briefly explain to students what you expect them to do in order to complete each choice. For example, you might establish that during the next three sessions each student should play Multiplication Pairs twice, each student should do three to four multiplication clusters, or each student should skip count with a partner and do three to four division problems. Tell students they may do more if they have time.

Choices 1 Through 4

For Choices 1 through 4, you can find a description of the activity in Choice Time: Multiplying, Dividing, and Skip Counting (p. 30). At this time, most students should be ready to use the Array Cards in Set B.

Choice 5: Multiplication Clusters

Students choose two or three multiplication clusters from Student Sheet 11 to solve and discuss with their partners. They talk about strategies for finding the answer to the final problem in each cluster. They should add any problems they thought of that helped them solve the last problem.

After solving several clusters, each student chooses one of the clusters to write about by explaining how he or she solved the final problem in the cluster and what other problems he or she used to help solve it. Partners may choose the same or different cluster sets to write about. Partners read each other's writing to see if it clearly states the reasoning used. They can make suggestions to each other about what could be added or clarified.

Choice 6: Solving Division Word Problems

Students work in pairs or small groups to solve division problems. Students may choose from problems found on Student Sheet 12, Problems About Our Class, and Student Sheet 13, Recycling Problems. For each problem, students should write how they solved the problem, or draw a picture. They should write an equation using division notation when it is helpful in solving the problem.

❖ **Tip for the Linguistically Diverse Classroom** Pair limited-English-proficient students with English-proficient students to do the problems on Student Sheets 12 and 13. Ask English-proficient students to make the words comprehensible to their partners by reading each problem aloud, and by pointing to objects in the classroom or drawing a picture as they read key words.

Making Choices Students may initially need help in making choices. Those students who have decided on their first choice can independently begin their work. Some students may need your help in deciding which choice they want or need to go to. It will probably take a few choice sessions before students are feeling more confident with these routines and expectations.

While Students Are Working on the Choices

Take a few minutes to go over homework for Session 1, helping students compare their solutions and strategies. If students have invented their own sets of cluster problems, let them share the sets at this time.

During Sessions 2, 3, and 4 you might want to work with a small group of students who are having difficulty with specific choice items. Enlist the help of students who understand choice items to act as "resources" or "helpers" for their peers who might have questions.

During today's Choice Time, I'm going to be working with a small group of students on Multiplication Clusters. If you think you would like to work on this choice or if you feel like you're not really understanding how to do Multiplication Clusters, come join us. If you have a question or need some help during Choice Time, you might ask a friend for some help. Is there anyone who feels that he or she understands skip counting, the array games, or Division Word Problems and would be willing to help someone who has a question?

Discussing Choice Activities At some point during these three sessions, have two class discussions. One discussion should focus on students sharing their strategies for solving a set of multiplication clusters (Set D, Student Sheet 11, page 2 of 5) and the other on a division word problem (Problem 2, Student Sheet 13). Both problems are from the student sheets they have been working on during Choice Time. Prior to each discussion, remind students to complete each of these problems. More detail about the discussions is given on pp. 52 and 53.

Class Discussion: Strategies for Solving Two-Digit Multiplication Problems

The focus of this discussion is for students to share their strategies for solving two-digit-by-one-digit multiplication problems. As students share their ideas, use the following questions to help you evaluate student work:

■ Are students using multiplication relationships they know to build solutions to harder problems?

■ Are students able to multiply by multiples of 10? Do they know why 20×7 is 140 or are they just "adding a zero" to 14?

■ Are students comfortably splitting multiplication problems into subproblems, then putting the solutions of the smaller problems back together?

■ Do students consistently visualize multiplication as groups of things? Do they use the array structure to then "see" multiplication relationships?

Put this multiplication cluster on the board or overhead:

$$5 \times 7$$
$$10 \times 7$$
$$4 \times 25$$
$$20 \times 7$$
$$25 \times 7$$

Here's a multiplication cluster you have been working on during Choice Time. What are some of the ways you solved the first four problems in the cluster?

Encourage students to think about which multiplication pairs or relationships they "just knew" versus those pairs where they used skip counting or some other strategy. Students don't need to elaborate on how they solved single-digit multiplication problems if they have become automatic.

Many students will see a pattern when a number is multiplied by 10. They may notice that you always add a zero and be unable to explain why. The **Teacher Note,** Multiplying by Multiples of 10 (p. 54) provides some ideas for discussing this pattern with students.

How did you use the first four problems to help you solve 25×7? Are there any other problems you added to the cluster to help you solve it?

Encourage students to talk about how they solved the final problem in the cluster, 25×7. There will probably be a variety of strategies, including pulling apart the problem into more manageable parts (10×7, 10×7, 5×7), skip counting by 25's, or thinking in terms of other key relationships (there are four 25's in 100). All of these ways are acceptable approaches.

Making Up a Cluster Problem   On the board or overhead, put up the problem 21 × 8 and ask students to make up a set of cluster problems that would help them to solve the problem. Encourage them to think about how they might break apart this larger problem into smaller, more familiar problems.

List their ideas on the board. If there is time, have students solve 21 × 8, using whichever problems in the cluster are helpful to them.

In this discussion, students talk about the solution to one of the division word problems they have been working on (Problem 2, Student Sheet 13). Put the following problem on the board or overhead:

Students have collected 90 cans for a recycling project. How many boxes do they need to store the cans? Each box holds 12 cans.

Have students demonstrate their strategies for solving this problem. Some students may need cubes, others may want to record their strategy on the board or overhead. In all cases, make sure students represent the problem using division notation in addition to explaining how they solved the problem.

Class Discussion: How Many Boxes Do You Need?

Sessions 2, 3, and 4 Follow-Up

Multiplication Clusters at Home   At the end of Session 3, assign students Student Sheet 14, Multiplication Clusters at Home.

More Array Games   After Session 4, students teach someone at home one of the Array Games (Multiplication Pairs, Count and Compare, Small Array/Big Array). If they have already played all three games at home, they choose which game they would like to play again. The game rules and Array Cards for the games should already be at home.

Homework

One of the mathematical "tricks" we all learned to help us with hard multiplication was the rule, "When you multiply by 10, just add a zero; when you multiply by 100, add two zeroes; and so on." Even if you don't teach this rule explicitly, students readily see the pattern of adding zeroes as they work on related multiplication problems, such as 2×7 and 20×7, and will invent the rule themselves. While this pattern is a very useful one, the mathematics behind it is fairly difficult for students at this age to grasp completely. If rules are simply memorized and used without thought, they are often misremembered or misapplied as problems become more complex.

Therefore, while we want to encourage students to use the patterns they've discovered, we also want them to double-check their work using other strategies and to be able to estimate that their solution is a reasonable one. For example, if a student solves 5×30 by saying "It's like 5 times 3 and add a zero, so it's 150," we would ask, "And how do you know 150 is a reasonable answer? Is there another way you can prove it? How do you know it's not 1500?"

By using arrays, students have a chance to see how multiplying by 10 and multiples of 10 works. Imagine the array for 8×1. There is one 8 in this array, or you can think of it as eight

1's. Similarly, imagine the array for 8×10. Looking at it one way, there are ten 8's; looking at it another way, there are eight 10's. For 8×100, there are one hundred 8's or eight 100's. Our number system is based on units based on powers of ten: ones, tens, hundreds, thousands, and so forth. Each time we multiply a number by one of these units, we get a multiple of that unit:

$8 \times 1 =$	8 ones =	8
$8 \times 10 =$	8 tens =	80
$8 \times 100 =$	8 hundreds =	800
$8 \times 1000 =$	8 thousands =	8000

It may look like we are just "adding zeroes." In fact, we are generating the same number of groups of larger and larger units: 8 ones, 8 tens, 8 hundreds, 8 thousands.

Multiplying by multiples of 10, 100, and 1000—for example, 4×20 or 8×30, 6×700, 2×6000—is a bit more complex, but is based on multiplication by 10, 100, and so on. When students imagine an array for 8×30 and relate it to the array for 8×10, they begin to see that they are multiplying *8 groups of three 10's* or $8 \times (3 \times 10)$. By visualizing and discussing what the arrays for such multiplication problems look like, they can begin to develop a sound basis for understanding the "add the zero" rule.

Problems That Look Hard But Aren't

What Happens

Materials

- Interlocking cubes (class set available)
- One-centimeter graph paper (1–2 per student)
- Student Sheet 15 (1 per student)
- Student Sheet 16 (1 per student, homework)
- Overhead projector, transparency pen

In the final session of this unit, students work in small groups to generate a list of "hard" multiplication pairs (or relationships). The class compiles a list of these pairs and discusses which are the most difficult relationships to learn and which they have developed strategies for solving. They solve a two-digit multiplication problem as a whole class. As an assessment activity, students solve another two-digit problem in two different ways. Their work focuses on:

- applying multiplication relationships to a two-digit problem
- using what they know about multiplication to discuss and solve harder problems

Activity

Discussing Multiplication Pairs

Organize students into groups of four. Each group is responsible for making a list of the five multiplication pairs that are the most difficult to learn (up to 12's table). You might suggest to students that they look back to the list they made when they played Multiplication Pairs. The group needs to agree on five pairs to present to the class. Groups will need 5 to 10 minutes to prepare their lists.

Activity

Which Pairs Are Hard for You?

Each group presents the list of multiplication pairs that its members consider the most difficult to learn. There may be pairs that are common to many groups' lists. As you record these pairs on the board, put a check mark next to any that are repeated (instead of recording them more than once). From this list determine a class list of the five to eight multiplication relationships that are the most difficult to learn. These will be the ones that have the most check marks next to them. Challenge students to learn to solve these problems in the next few days.

There may be some difference of opinion about pairs some students consider hard and others do not. Have students share strategies for how they learned or remember these relationships. Some strategies students have used in the past for learning hard relationships are breaking the problem into two easier problems, relating the problem to one they already know, or picking one or two hard problems each day to learn. The **Dialogue Box**, Strategies for Learning "Hard Problems," p. 59, gives some examples of student strategies.

Problems That Look Hard But Aren't

Present a two-digit by one-digit multiplication problem such as this on the board or overhead:

$$31 \times 6$$

Some people think this looks like a hard multiplication problem. During the past two weeks you've been developing strategies for solving these kinds of problems. I'd like you to talk briefly with the person next to you about how you would solve this problem.

Students talk with the people near them about how they would solve the problem. Bring the group together to share strategies for approaching this multiplication problem. Record students' ideas on the board. Ask students why some people might think this was a hard multiplication problem.

What are some other two-digit multiplication problems that look hard but really aren't?

Record this list on the board and save it. Students will choose one problem from the list to do for homework.

Assessment

Solving Two-Digit Multiplication Problems

Students work alone to solve the next problem. They solve the problem in two different ways and write about how they solved the problem. Graph paper and interlocking cubes should be available for those students who need them.

Pass out a copy of Student Sheet 15, Two Ways to Solve a Problem (p. 111), to each student. Tell students you would like to get an idea of how they are thinking about multiplication problems they have been doing and what kinds of strategies they are using.

These responses are one way of assessing a student's understanding of multiplication problems. See the **Teacher Note**, Two Ways to Solve 27×4 (p. 58). As you look at student papers, there are several questions you can focus on:

■ Are students keeping track of their work carefully?

■ Are students becoming fluent in the basic multiplication relationships?

■ Do students demonstrate an understanding of partitioning large numbers into more familiar parts as a way of multiplying?

■ Are students using tools (centimeter paper, cubes) appropriately to help them solve the problem?

Choosing Student Work to Save

As the unit ends, you may want to use one of the following options for creating a record of students' work on this unit:

- Students look back through their folders or notebooks and write about what they learned in this unit, what they remember most, and what was hard or easy for them. Students could do this during their writing time.

- Students select one or two pieces of their work as their best, and you also choose one or two pieces of their work to be saved. This work may be saved in a portfolio for the year. You might include students' written solutions to the assessment, Creating and Solving Division Problems (Investigation 2, Sessions 7 and 8), and any other assessment tasks from this unit. Students can create a separate page with brief comments describing the pieces of work.

- You may want to send a selection of work home for parents to see. Students write a cover letter describing their work in this unit. This work should be returned if you are keeping a year-long portfolio of mathematics work for each student.

Session 5 Follow-Up

Most likely, all students will not have had the opportunity to finish every choice item. You might decide to have these choices remain in the classroom for students to work on during their free time or for homework.

Note: Students will need to use their sets of arrays later in the year in the unit *Packages and Groups*. We suggest students keep their bags of arrays in their current math folders and continue to use them as a way of learning the basic multiplication relationships. You will also need to save array transparencies.

Problems That Look Hard But Aren't On Student Sheet 16, students choose one multiplication problem from the class list of "Problems That Look Hard But Aren't" to solve and write about.

Two Ways to Solve 27 × 4

As you look at students' work for this assessment, consider whether each student has developed reliable strategies and can record clearly the steps used to solve the problem. Has the student used a combination of addition and multiplication strategies? Can the student pull apart a "hard" multiplication problem into smaller related multiplication problems? Asking students to solve problems in more than one way helps them to think flexibly and also gives them a way to check their work. The following are examples of students' solutions to the problem 27 × 4.

■ *Addition Strategies.* Many students see this problem as one of repeated addition.

DeShane

$$
\begin{array}{r}
27 \\
27 \\
27 \\
+\ 27 \\
\hline
108
\end{array}
$$

I added the 7's and got 28. Then I added the 20's and got 80. So 80 + 28 = 108

Some students used the same notation but explained that they "skip counted by 27's."

Teresa

$$
\begin{array}{r}
27 \\
27 \\
27 \\
+\ 27 \\
\hline
108
\end{array}
$$

I skip counted by 27s

Still others may know that if two 27's is equal to 54 then four 27's is double that. One student represented the problem in this way:

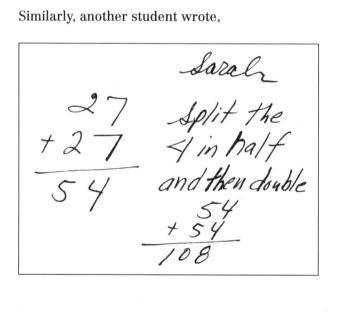

Qi Sun

I know 27 + 27 = 54 and so 4 27s is double 54. So,

$$
\begin{array}{r}
27 + 27 = 54 \\
+\ 54 \\
\hline
108
\end{array}
$$

Similarly, another student wrote,

Sarah

$$
\begin{array}{r}
27 \\
+\ 27 \\
\hline
54
\end{array}
$$

Split the 4 in half and then double

$$
\begin{array}{r}
54 \\
+\ 54 \\
\hline
108
\end{array}
$$

Continued on next page

■ *Multiplication Strategies.* We expect that many students will be able to break apart the problem into smaller, more familiar multiplication problems as one of their strategies. Here is how one student kept track of each part of the problem and of the totals of those problems:

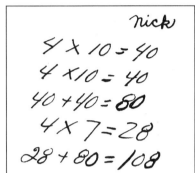

Some students will break 27 into 20 + 7, multiply each by 4, and add the subtotals:

> "I counted 20 four times and got 80. Then I said 7 × 4 and got 28, and then I added them together and got 108."

Here is another version of a student's record of this solution:

Rashaida
20 × 4 = 80
7 × 4 = 28
80 + 28 = 108

Strategies for Learning "Hard Problems"

This discussion occurred while students were working on the activity Which Pairs Are Hard for You? (p. 55).

DeShane: I think 6 × 8 is a hard problem. It's on my list every time.

[Lots of students groan and agree with DeShane. A few students in the class disagree.]

Teresa: I had 6 × 8 on my list too, but you can think of it like 6 × 4 and 6 × 4. That's 24 plus 24, and that's easier to figure out for me. It's like the Small Array/Big Array game. 6 × 4 is the small array and 6 × 8 is the big array, so it's a match.

Rashaida: I just think of 5 × 8 which I know is equal to 40, then I add 8 more.

DeShane: That's hard.

Qi Sun: I just memorized it. I kept thinking, 6 × 8 is equal to 48, 6 × 8 is equal to 48; then I just remembered it.

Those are some interesting strategies for remembering a hard problem. Teresa broke the problem down into pieces she knew. Rashaida connected it to a fact she already knew and then added on, and Qi Sun practiced it over and over again. Any other ideas? How about skip counting?,

Sarah: You could skip count by 8's, but that's tricky.

DeShane: Or you could do 6's, that's tricky too. I think I'll try skip counting by 6's with my partner today at Choice Time.

Sounds like a good idea.

Counting Around the Class

Basic Activity

Students count around the class by a particular number. That is, if counting by 2's, the first student says "2," the next student says "4," the next "6," and so forth. Before the count starts, students try to predict on what number the count will end. During and after the count, students discuss relationships between the chosen factor and its multiples.

Counting Around the Class is designed to give students practice with counting by many different numbers and to foster numerical reasoning about the relationships among factors and their multiples. Students focus on:

- becoming familiar with multiplication patterns
- relating factors to their multiples
- developing number sense about multiplication and division relationships

Materials

- Calculators (for variation)

Procedure

Step 1. Choose a number to count by. For example, if the class has been working with quarters recently, you might want to count by 25's.

Step 2. Ask students to predict the target number. "If we count by 25's around the class, what number will we end up on?" Encourage students to talk about how they could figure this out without doing the actual counting.

Step 3. Count around the class by your chosen number. "25 . . . 50 . . . 75 . . ." If some students seem uncertain about what comes next, you might put the numbers on the board as they count; seeing the visual patterns can help some students with the spoken one.

You might count around a second time by the same number, starting with a different person, so that students will hear the pattern more than once and have their turns at different points in the sequence.

Step 4. Pause in the middle of the count to look back. "We're up to 375 counting by 25's. How many students have we counted so far? How do you know?"

Step 5. Extend the problem. Ask questions like these:

"Which of your predictions were reasonable? Which were possible? Which were impossible?" (A student might remark, for example, "You couldn't have 510 for 25's because 25 only lands on the 25's, the 50's, the 75's, and the 100's.")

"What if we had 32 students in this class instead of 28? Then where would we end up?"

"What if we used a different number? This time we counted by 25's and ended on 700; what if we counted by 50's? What number do you think we would end on? Why do you think it will be twice as big? How did you figure that out?"

Variations

Multiplication Practice Use single-digit numbers to provide practice with multiplication (that is, count by 2's, 3's, 4's, 5's, 6's, and so forth). In counting by numbers other than 1, students usually first become comfortable with 2's, 5's, and 10's, which have very regular patterns. Soon they can begin to count by more difficult single-digit numbers: 3, 4, 6, and (later) 7, 8, and 9.

Landmark Numbers When students are learning about money or about our base ten system of numeration, they can count by 20's, 25's, 50's, 100's, and 1000's. Counting by multiples of 10 and 100 (30's, 40's, 600's) will support students' growing familiarity with the base 10 system of numeration.

Continued on next page

Making Connections When you choose harder numbers, pick those that are related in some way to numbers students are very familiar with. For example, once students are comfortable counting by 25's, have them count by 75's. Ask students how knowing the 25's will help them count by 75's. If students are fluent with 3's, try counting by 6's or by 30's. If students are fluent with 10's and 20's, start working on 15's. If they are comfortable counting by 15's, ask them to count by 150's or 1500's.

Large Numbers Introduce large numbers, such as 2000, 5000, 1500, or 10,000, so that students begin to work with combinations of these less familiar numbers.

Using the Calculator On some days you might have everyone use a calculator, or have a few students use the calculators to skip count while you are counting around the class. On most calculators, the equals (=) key provides a built-in constant function, allowing you to skip count easily. For example, if you want to skip count by 25's, you press your starting number (let's say 0), the operation you want to use (in this case, +), and the number you want to count by (in this case, 25). Then, press the equals key each time you want to add 25. So, if you press

you will see on your screen 25, 50, 75, 100.

Counting Patterns Students write out a counting pattern up to a target number (for example, by 25's up to 500). Then they write about what patterns they see in their counting. Calculators can be used for this.

Mystery Number Problems Provide an ending number and ask students to figure out what factor they would have to count by to reach it. For example: "I'm thinking of a mystery number. I figured out that if we counted around the class by my mystery number today, we would get to 2800. What is the mystery number?"

Or, you might provide students with the final number and the factor, and ask them to figure out the number of students in the class. "When a certain class counts by 25's, the last student says 550. How many students are in the class?" Calculators can be used.

Special Notes

Letting Students Prepare When introducing an unfamiliar number to count with, students may need some preparation before they try to count around the class. Ask students to work in pairs to figure out, with whatever materials they want to use, what number the count will end on.

Avoiding Competition It is important to be sensitive to potential embarrassment or competition if some students have difficulty figuring out their number. One teacher allowed students to volunteer for the next number, rather than counting in a particular order. Other teachers have made the count a cooperative effort, establishing an atmosphere in which students readily helped each other, and anyone felt free to ask for help.

Multiple BINGO

Basic Activity

This game can be played either as a whole class or with a partner or a small group. Students can play it independently, but it's more fun to play with other people. The object of the game is to color five numbers in a row, either across, up and down, or diagonally. The factor cards tell you what numbers you can color in. The number you color in must be a multiple of the factor card.

Multiple BINGO is designed to give students practice with finding multiples of numbers. Either by skip counting or using multiplication pairs, this activity fosters numerical reasoning about the relationships among factors and their multiples. Students focus on:

- relating factors to their multiples
- becoming familiar with multiplication patterns
- developing number sense about multiplication and division relationships.

Materials

- 100 chart (one for each player)
- Deck of factor cards (one deck per 2–3 students)
- Crayons or markers
- Optional: Students may use 100 charts on which they have highlighted 2's, 3's, 4's, 5's, and so on, or they may use calculators.

Procedure

Playing as partners or in a small group

Step 1. Gather materials. Each player has a 100 chart. Put the factor cards in a pile in the middle of the table. Players take turns picking a factor card for the group.

Step 2. Choose a multiple. Every player colors in *one* number that is a multiple of the factor

card and writes the factor in the square. For example, if someone turns over a 5 card, any one of the numbers 5, 10, 15, 20, 25, and so on can be chosen. The next player picks a factor card, and again every player chooses *one* number to color in.

Step 3. Using a Wild Card. If a Wild Card is picked, the player who picked it can decide on the factor to be used. Any number from 1 to 100 can be chosen when a Wild Card is drawn. For game strategy, the player should choose a number that helps his or her game but doesn't help the other players. Often the most useful number to pick is a prime number, such as 23, to fill in a gap between other multiples; other players could mark 23, or 46, or 69, or 92.

Step 4. Color five multiples in a row. The game continues until a player colors five numbers in a row and gets BINGO. Players can choose to continue until other players also get five in a row.

Variations

Whole Class Game This game can be played as a whole class. A leader, the teacher or a student, draws the factor cards for the whole class. If a Wild Card is drawn, the leader calls on a player to choose a factor for the group to use. The object of the game is to cover five numbers in a row. You might want to continue play until every player has covered five in a row. When the class plays as a large group, there is the option for students who are new to the game to collaborate with another student.

Limiting the Factors An easier version of Multiple BINGO is to use only the 2's, 3's, 4's, and 5's factor cards and a few Wild Cards. Students can use the 100 charts they have highlighted for the multiples of 2, 3, 4, and 5 as references while they play the game. As students highlight 100 charts for new multiples, add those new factor cards to the game.

Continued on next page

Limiting the 100 Chart When students first play Multiple BINGO, they will tend to use only "easy" numbers—especially the single-digit numbers and the multiples of 10. To encourage them to use more difficult numbers, you might:

1. Have them omit the top row and right column of the 100 chart.

2. Insist they start with a number near the middle of the chart.

3. Give two points for a win that is a diagonal (five numbers next to each other on any diagonal is fine). This may encourage them to notice the 9's and 11's tables on the two main diagonals.

Other Ways to Win Once students are familiar with the basic game, vary the ways to win. For example, a "win" could be defined as coloring six multiples in a row or coloring in a 2 × 3 rectangle of multiples.

Special Notes

Using Resources Encourage students to use calculators or their sets of highlighted 100 charts they made for the multiples 2 to 12 as a way of determining multiples for the factor cards.

Reusing the 100 Chart Students can use a contrasting color to play another game on the same 100 chart. Some teachers have laminated a set of 100 charts that can be wiped off after each game. Students will need to use either a crayon or an overhead marker with these laminated charts.

The following activities will help ensure that this unit is comprehensible to students who are acquiring English as a second language. The suggested approach is based on *The Natural Approach: Language Acquisition in the Classroom* by Stephen D. Krashen and Tracy D. Terrell (Alemany Press, 1983). The intent is for second-language learners to acquire new vocabulary in an active, meaningful context.

Note that *acquiring* a word is different from *learning* a word. Depending on their level of proficiency, students may be able to comprehend a word upon hearing it during an investigation without being able to say it. Other students may be able to use the word orally but not read or write it. The goal is to help students naturally acquire targeted vocabulary at their present level of proficiency.

We suggest using these activities just before the related investigations. The activities can also be led by English-proficient students.

Investigations 1–3

packages packaged

1. Display a carton of eggs, a bag of pretzels, and a box of pencils.

2. Explain that *packaging* something means to put items together.

3. Point out how the displayed items are packaged differently.

   The eggs are packaged in a cardboard carton. The pretzels are packaged in a bag. The pencils are packaged inside a box.

4. Point out how some items are arranged in rows and others are not.

5. Ask students to decide other ways eggs, pretzels, and pencils could or could not be packaged.

   Would you package eggs inside a plastic bag? Could you package pretzels inside a box? Why? Why not?

related

1. Show students three clear plastic bags with the following items placed inside:

   Bag 1: drawing materials such as: oil pastels, marker pens, and crayons;

   Bag 2: items of all one color

   Bag 3: paper clips, rubber bands, string

2. Identify the items in Bag 1. Explain that although each item in the bag is different, they are actually all related to each other. Further explain that this means there is something the same about all three items. Ask yes or no questions to help students identify the relationship of these items.

   Are the items related to each other because they are all the same size? Are the items related to each other because they are all made out of the same material? Are the items related to each other because they are all the same color? Are the items related to each other because they are all used for drawing?

3. Continue with the same format for Bags 2 and 3.

4. Challenge students' comprehension of the word *related* by playing the following game. Have students collect various items from home and around the classroom.

   Tell them to put three related items into a plastic bag. Then have them put in one more item that is not related to the other items. Have students share their bags. Challenge the other students in the group to identify which item in each bag is not related and why.

strategy, strategies

1. Display 12 interlocking cubes in front of the students.

2. Ask them to think of a fast way to count the number of cubes in this group.

3. Point out that there are several different ways to do this task as you show examples of different strategies. Be sure to articulate your actions as you are doing them. Emphasize that there is not just one "correct" strategy.

   One strategy might be to divide the cubes in half, count one of these groups, and then double that number. Another strategy or idea might be to count the cubes by 2's, 3's, or 4's.

4. Next, pair students and have them sit around a large pile of interlocking cubes. Challenge them to think of a strategy for using the cubes to make three towers as quickly as possible. Tell them each tower needs to have six cubes.

5. When students have completed the task, ask questions about their individual strategies.

   Did you both work on building each tower? Did that strategy help you finish quickly? Did one person count and another build? Did that strategy help you finish quickly? Did you count by 2's? Did that strategy help you finish quickly?

Blackline Masters

Dear Family,

In mathematics, our class is starting a new unit called *Arrays and Shares*. This unit focuses on multiplication and division. Students begin the unit by looking at things that are arranged in rows, for example, juice packs, egg cartons, and rows of chairs. Through examining these rectangular arrangements (or arrays), they begin to visualize important aspects of multiplication, for example, that the solution to 7×6 is the same as the solution to 6×7.

As students work on two-digit multiplication and related division problems, it is critical that they visualize how to pull apart numbers they are working with. To solve harder problems, students learn to use related problems they already know how to solve. For example, the problem 7×23 can be solved by breaking the problem into more familiar parts: 7×10, 7×10, and 7×3.

While our class is studying multiplication and division, you can help in the following ways:

> Look for items around your house or at the grocery store that are packaged or arranged in rectangular arrays: tiles on the floor, egg cartons, window panes, six-packs of juice cans, and the like. Talk with your child about the dimensions (rows and columns) and discuss ways to figure out the total number.

> Play the Array Games that your child brings home for homework.

> Help your child practice skip counting by 3's, 4's, 5's, and so forth.

Emphasis during this unit will be on thinking hard and reasoning carefully to solve mathematical problems. Students are encouraged to develop strategies that make sense to them based on understanding numbers and their relationships, so that they can solve problems easily and flexibly. They are expected to develop more than one way to solve a problem so that they can use one method to double-check another. Some of these methods may not be the ones you learned in school, although you may recognize some of them as methods you use in your daily life.

When your child brings home problems, encourage your child to explain his or her strategies to you. Ask questions, such as "How did you figure that out?" and "Tell me your thinking about this problem," but don't provide answers or methods. Show that you are interested in how your child is thinking and reasoning about these problems.

Thank you for your interest in your child's study of mathematics. We are looking forward to an exciting few weeks of work on multiplication and division.

Sincerely,

Multiples on the 100 Chart

I am completing the multiples of _____

1	2	3	4	5	6	7	8	9	10
11	12	13	14	15	16	17	18	19	20
21	22	23	24	25	26	27	28	29	30
31	32	33	34	35	36	37	38	39	40
41	42	43	44	45	46	47	48	49	50
51	52	53	54	55	56	57	58	59	60
61	62	63	64	65	66	67	68	69	70
71	72	73	74	75	76	77	78	79	80
81	82	83	84	85	86	87	88	89	90
91	92	93	94	95	96	97	98	99	100

Pattern observed _____

_____

I am completing the multiples of _____

1	2	3	4	5	6	7	8	9	10
11	12	13	14	15	16	17	18	19	20
21	22	23	24	25	26	27	28	29	30
31	32	33	34	35	36	37	38	39	40
41	42	43	44	45	46	47	48	49	50
51	52	53	54	55	56	57	58	59	60
61	62	63	64	65	66	67	68	69	70
71	72	73	74	75	76	77	78	79	80
81	82	83	84	85	86	87	88	89	90
91	92	93	94	95	96	97	98	99	100

Pattern observed _____

_____

I am completing the multiples of _____

1	2	3	4	5	6	7	8	9	10
11	12	13	14	15	16	17	18	19	20
21	22	23	24	25	26	27	28	29	30
31	32	33	34	35	36	37	38	39	40
41	42	43	44	45	46	47	48	49	50
51	52	53	54	55	56	57	58	59	60
61	62	63	64	65	66	67	68	69	70
71	72	73	74	75	76	77	78	79	80
81	82	83	84	85	86	87	88	89	90
91	92	93	94	95	96	97	98	99	100

Pattern observed _____

_____

I am completing the multiples of _____

1	2	3	4	5	6	7	8	9	10
11	12	13	14	15	16	17	18	19	20
21	22	23	24	25	26	27	28	29	30
31	32	33	34	35	36	37	38	39	40
41	42	43	44	45	46	47	48	49	50
51	52	53	54	55	56	57	58	59	60
61	62	63	64	65	66	67	68	69	70
71	72	73	74	75	76	77	78	79	80
81	82	83	84	85	86	87	88	89	90
91	92	93	94	95	96	97	98	99	100

Pattern observed _____

_____

Multiplication Clusters

Solve the first two problems in each cluster. Then use those problems to help you solve the last problem. If you think of another problem that helps you solve the last problem, add it to the cluster.

4×4 $\begin{array}{r} 8 \\ \times\,4 \\ \hline \end{array}$ $\begin{array}{r} 12 \\ \times\,4 \\ \hline \end{array}$

7×2 $\begin{array}{r} 8 \\ \times\,2 \\ \hline \end{array}$ 9×2

5×6 6×6 $\begin{array}{r} 7 \\ \times\,6 \\ \hline \end{array}$

3×5 $\begin{array}{r} 10 \\ \times\,5 \\ \hline \end{array}$ $\begin{array}{r} 12 \\ \times\,5 \\ \hline \end{array}$

$\begin{array}{r} 9 \\ \times\,3 \\ \hline \end{array}$ $\begin{array}{r} 9 \\ \times\,6 \\ \hline \end{array}$ 9×12

Investigation 1 • Session 3
Arrays and Shares

More Multiplication Clusters

Solve the first two problems in each cluster. Then use those problems to help you solve the last problem. If you think of another problem that helps you solve the last problem, add it to the cluster.

8 × 2	8 × 3	8 × 5

5 × 2	10 × 2	15 × 2

7 × 2	7 × 4	7 × 8

3 × 6	6 × 6	12 × 6

12 × 5	10 × 7	12 × 7

1	2	3	4	5	6	7	8	9	10
11	12	13	14	15	16	17	18	19	20
21	22	23	24	25	26	27	28	29	30
31	32	33	34	35	36	37	38	39	40
41	42	43	44	45	46	47	48	49	50
51	52	53	54	55	56	57	58	59	60
61	62	63	64	65	66	67	68	69	70
71	72	73	74	75	76	77	78	79	80
81	82	83	84	85	86	87	88	89	90
91	92	93	94	95	96	97	98	99	100

Investigation 1 • Resource
Arrays and Shares

I am completing the multiples of _____

1	2	3	4	5	6	7	8	9	10
11	12	13	14	15	16	17	18	19	20
21	22	23	24	25	26	27	28	29	30
31	32	33	34	35	36	37	38	39	40
41	42	43	44	45	46	47	48	49	50
51	52	53	54	55	56	57	58	59	60
61	62	63	64	65	66	67	68	69	70
71	72	73	74	75	76	77	78	79	80
81	82	83	84	85	86	87	88	89	90
91	92	93	94	95	96	97	98	99	100

Pattern observed _____

_____

I am completing the multiples of _____

1	2	3	4	5	6	7	8	9	10
11	12	13	14	15	16	17	18	19	20
21	22	23	24	25	26	27	28	29	30
31	32	33	34	35	36	37	38	39	40
41	42	43	44	45	46	47	48	49	50
51	52	53	54	55	56	57	58	59	60
61	62	63	64	65	66	67	68	69	70
71	72	73	74	75	76	77	78	79	80
81	82	83	84	85	86	87	88	89	90
91	92	93	94	95	96	97	98	99	100

Pattern observed _____

_____

I am completing the multiples of _____

1	2	3	4	5	6	7	8	9	10
11	12	13	14	15	16	17	18	19	20
21	22	23	24	25	26	27	28	29	30
31	32	33	34	35	36	37	38	39	40
41	42	43	44	45	46	47	48	49	50
51	52	53	54	55	56	57	58	59	60
61	62	63	64	65	66	67	68	69	70
71	72	73	74	75	76	77	78	79	80
81	82	83	84	85	86	87	88	89	90
91	92	93	94	95	96	97	98	99	100

Pattern observed _____

_____

I am completing the multiples of _____

1	2	3	4	5	6	7	8	9	10
11	12	13	14	15	16	17	18	19	20
21	22	23	24	25	26	27	28	29	30
31	32	33	34	35	36	37	38	39	40
41	42	43	44	45	46	47	48	49	50
51	52	53	54	55	56	57	58	59	60
61	62	63	64	65	66	67	68	69	70
71	72	73	74	75	76	77	78	79	80
81	82	83	84	85	86	87	88	89	90
91	92	93	94	95	96	97	98	99	100

Pattern observed _____

_____

Things That Come in Arrays

Item	Total	Dimensions	Draw the Array

Arranging Chairs

1. The kindergarten class is coming to watch a play in our classroom. There are 20 students. In what different ways could we arrange the chairs for them so that all the rows are equal? Use graph paper to create all the possible arrays, write the dimensions on each, and attach them to the back of this sheet.

2. The two third grade classes are going to watch our play in the cafeteria. There are 49 students all together. In what different ways could we arrange the chairs for them so that all the rows are equal? Use graph paper to create all the possible arrays, write the dimensions on each, and attach them to the back of this sheet.

3. What do you notice about your solutions for Problem 1 and Problem 2?

Pairs I Know, Pairs I Don't Know

Pairs I Know:

Pairs I Don't Know:

What Do You Do with the Extras?

1. There are 36 people who are taking a trip in some small vans. Each van holds 8 people. How many vans will they need?

2. Eight people are going to share 36 crackers equally. How many crackers will each person get?

3. Eight people are going to share 36 balloons equally. How many balloons will each person get?

4. There are 36 students who are going to see a movie. Each row holds 8 people. How many rows will they fill up?

5. Eight friends raised 36 dollars by washing people's cars. They want to share the money equally. How much should each person get?

What's the Story?

Write a story for each division problem.
Then solve the problem.

1. $42 \div 7$

2. $8 \overline{)52}$

3. $9 \overline{)63}$

4. $49 \div 4$

Word Problems

Solve each word problem. Write your solution in words. Then use division notation to write an equation that represents the problem.

1. Juice comes in 6-packs. How many 6-packs would we need to buy if we bought a can of juice for everyone in our class?

2. There are 46 people attending a concert. The chairs are set up in rows of 7. How many rows will be filled?

3. The book you are reading has 17 chapters. If you read 2 chapters in 3 hours, how many hours will it take you to read the book?

4. There will be 9 people at my pizza party. If each person eats 3 slices of pizza and each pizza has 8 slices, how many pizzas should I order?

THREE-QUARTER-INCH GRAPH PAPER

80

What You Will Need

- Copies of Array Card pages (Set A, pages 1–6; Set B, pages 7–17)
- Scissors, pencil, marker
- Plastic bags

What to Do

1. Carefully cut out each array on the Array Card pages. Follow the outlines of each array as exactly as you can.

2. Each card has two sides. On the array side of the card, write in pencil the **dimensions** of the array.
   Example:

   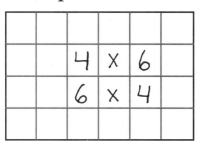

3. On the other side (the blank side), write in pencil the total number of squares in the array.
   Example:

   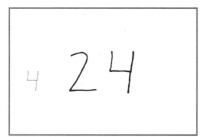

   Also write one of the dimensions of the array along one side. Write very lightly in pencil so that students can erase this dimension later.

4. Put each student set of cards in a separate plastic bag. Write the appropriate student name or names on each bag with permanent markers.

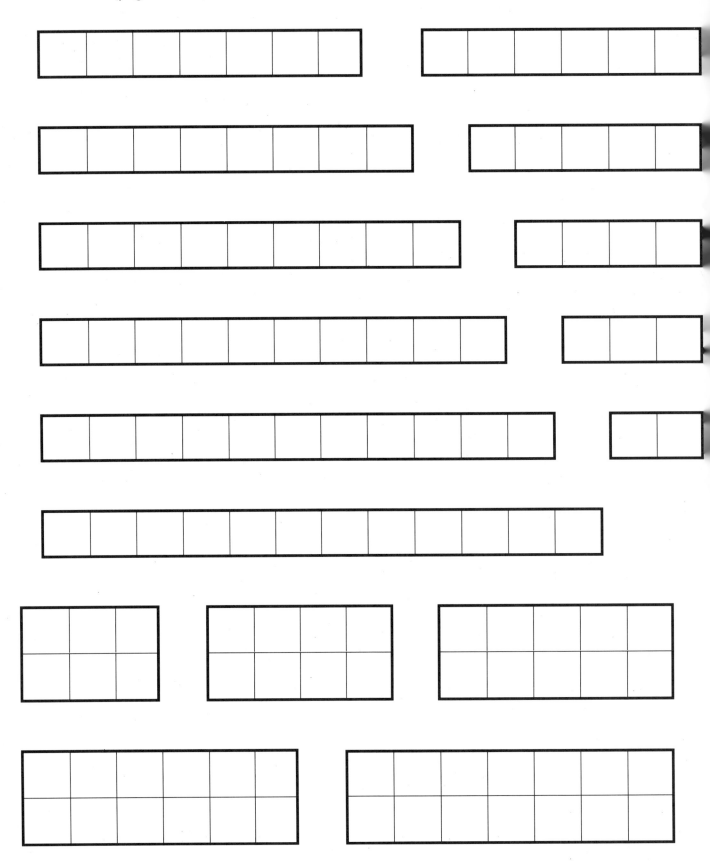

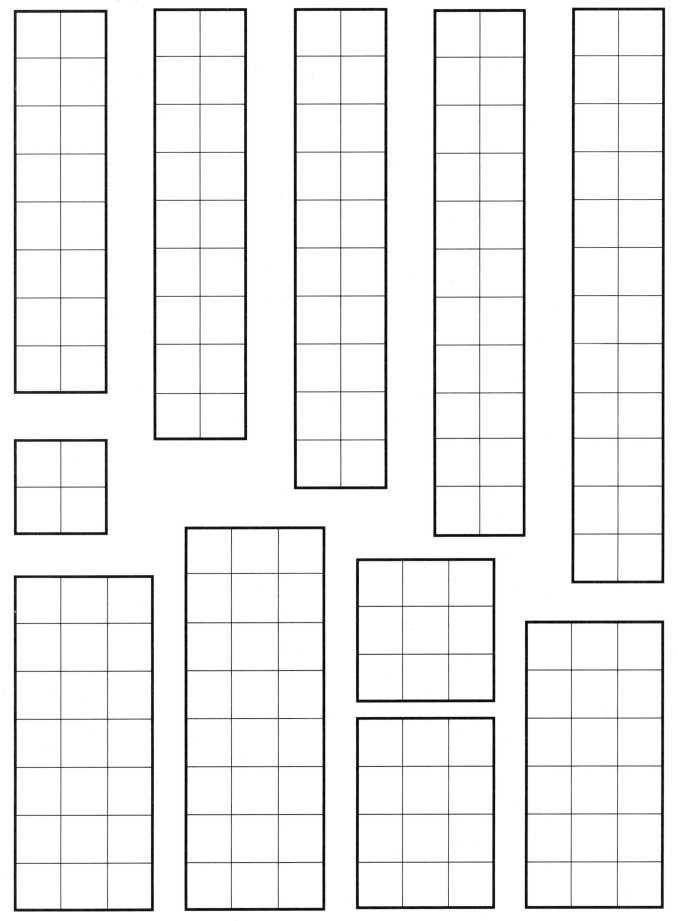

Investigation 2 • Resource
Arrays and Shares

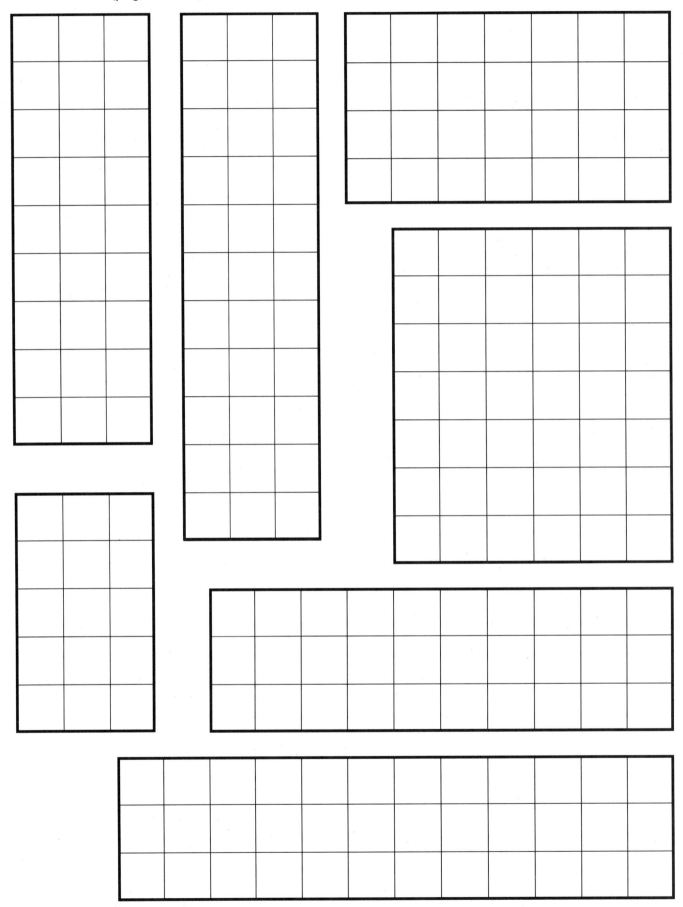

84

85

Investigation 2 • Resource
Arrays and Shares

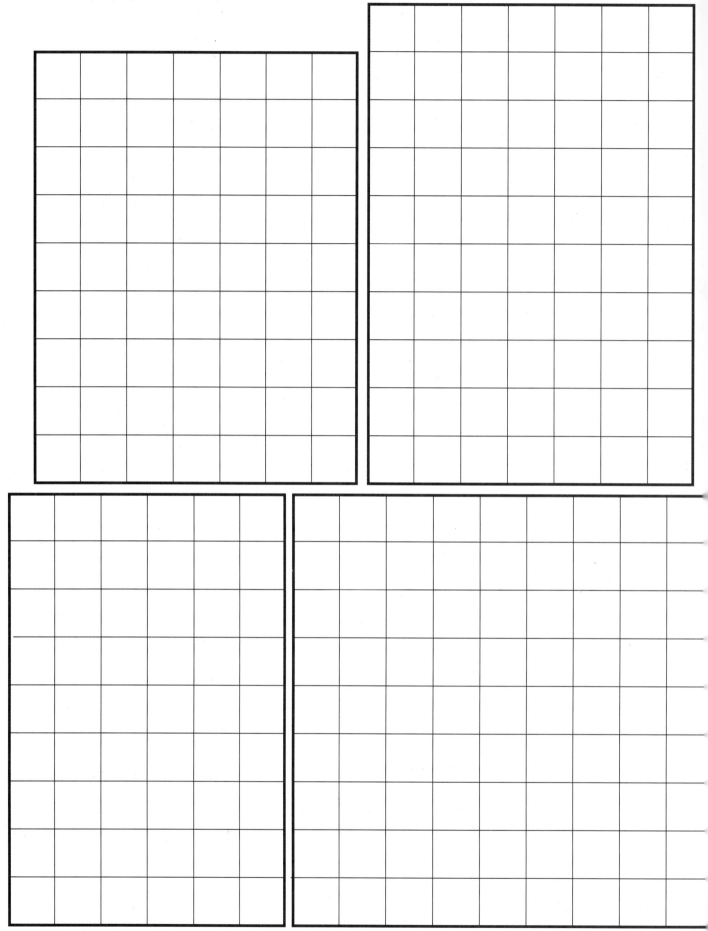

Investigation 2 • Resource
Arrays and Shares

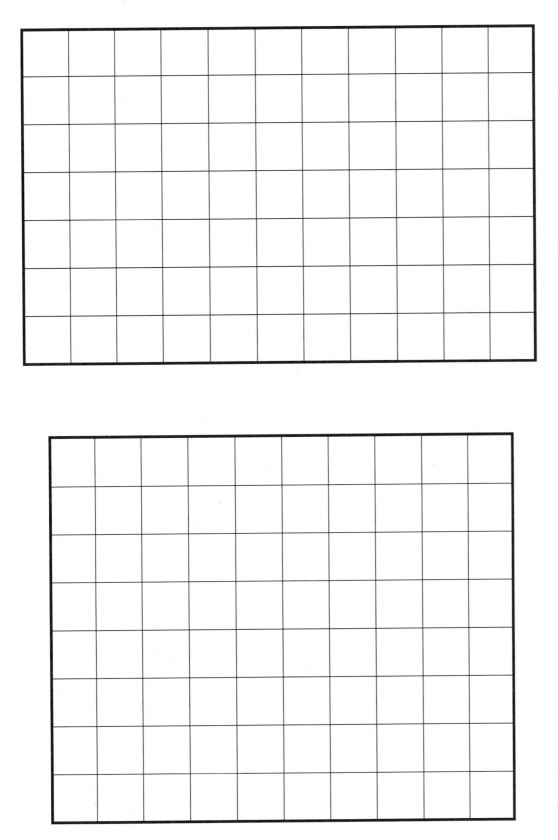

Investigation 2 • Resource
Arrays and Shares

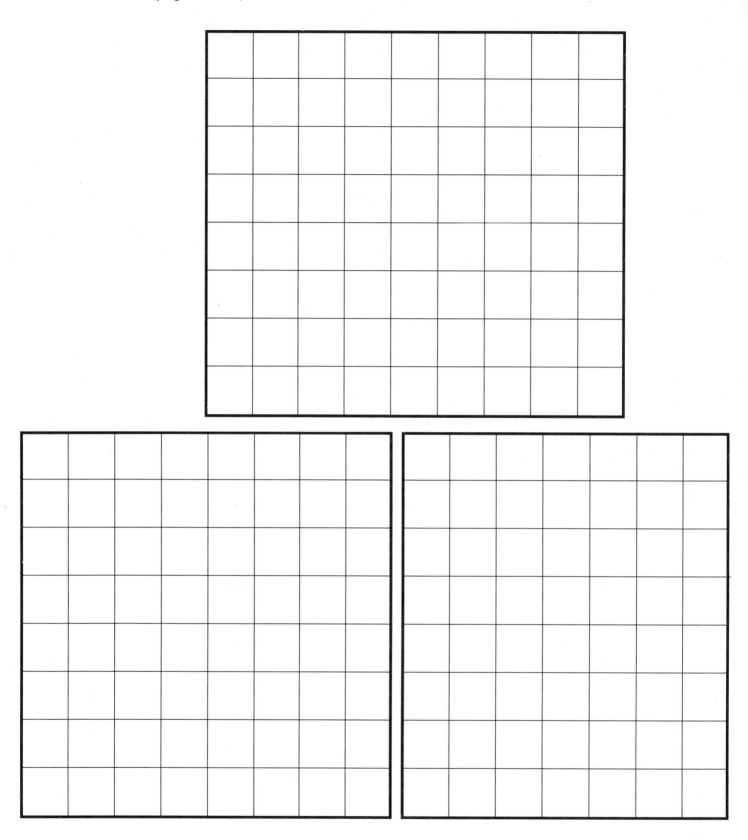

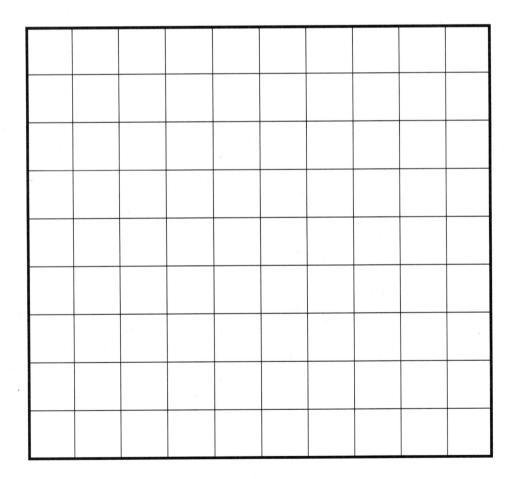

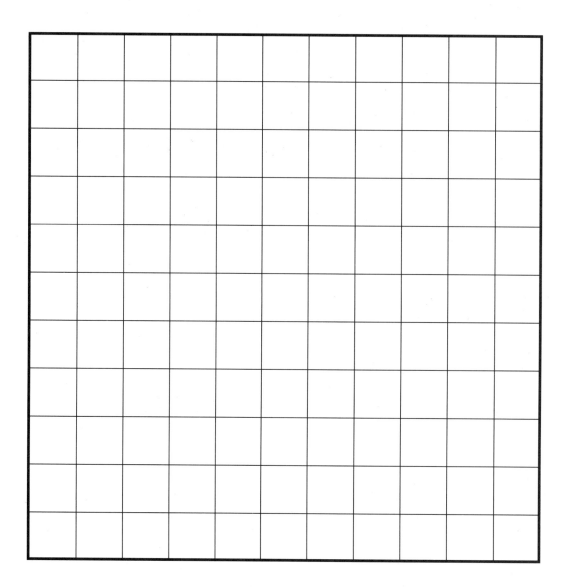

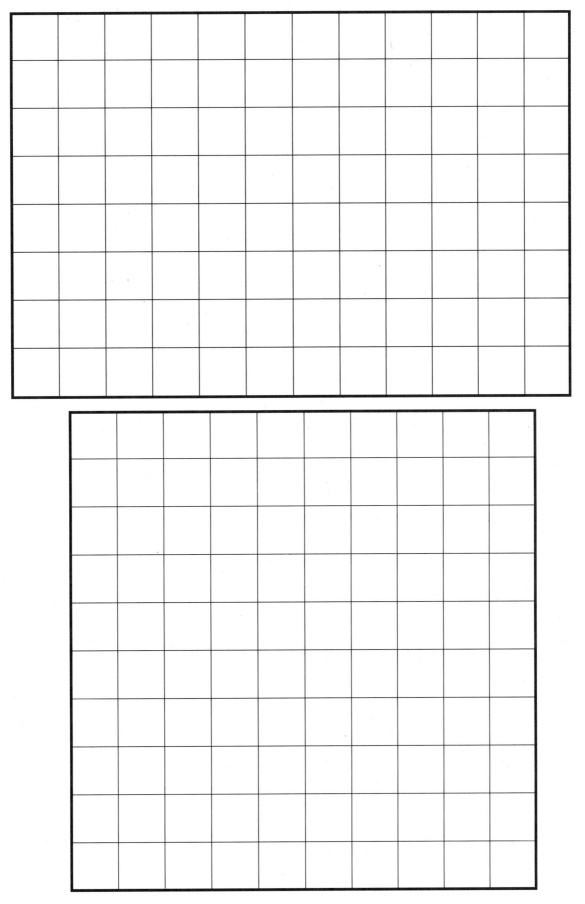

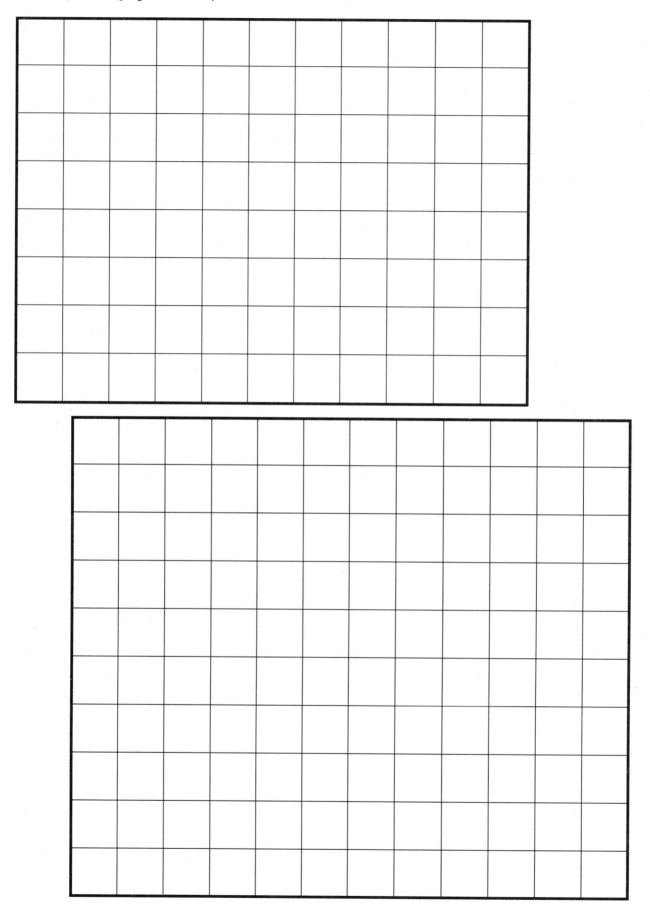

94

96

Investigation 2 • Resource
Arrays and Shares

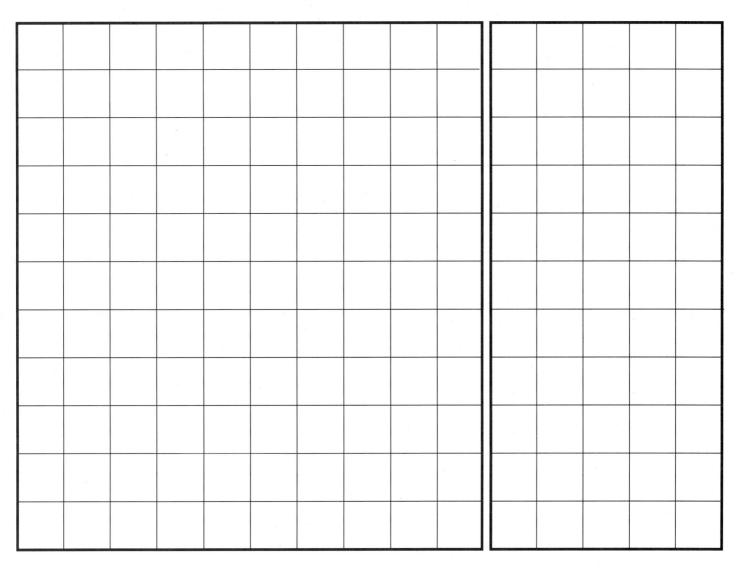

Investigation 2 • Resource
Arrays and Shares

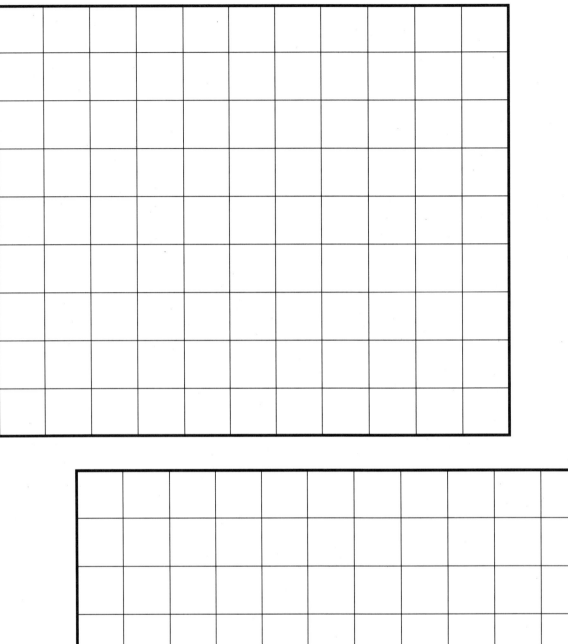

98

Materials

■ Set of Array Cards
■ Paper and pencil

Players: 1 or 2

How to Play

1. Spread out all the Array Cards in front of you. Some should be turned up, showing the dimensions. Others should be turned over to show the total.

2. Choose an Array Card and put your finger on it. (Don't pick it up until you say the answer.) If the dimensions are showing, you must give the total. If the total is showing, you must say the dimensions of the grid. The shape of the array will help you.

   For example: Suppose you pick an array with the total 36 showing. The dimensions could be 6×6, or 9×4, or 12×3. You must decide which is right. The shape of the array is a good clue.

3. Turn the card over to check your answer. If your answer is correct, then pick up the card.

4. If you are playing with a partner, take turns choosing and identifying cards. Play until you have picked up all the cards.

   While you are playing, make lists for yourself of "pairs that I know" and "pairs that I don't know yet." Use these lists to help you learn all the pairs.

Materials: Set of Array Cards

Players: 2 or 3

How to Play

1. If you are playing with a partner, sit across from each other. If three people are playing, sit in a circle.

2. Deal out the Array Cards with the total sides face down. Players should all have the same number of cards. Set aside any that are left over.

3. Place your cards in a stack in front of you, with the total side face down.

4. Players take the top card from their stacks and place these cards side by side (total sides still face down).

5. Decide which array is largest. You can do this just by looking, or by skip counting by rows to find the total of each. Counting the squares by 1's is not allowed.

6. The player with the largest array takes the cards, after proving that it is the largest.

7. Sometimes arrays with the same total may be played in one turn—like this:

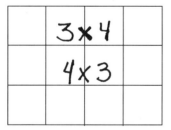 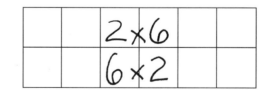

   When this happens, the players decide together who will get the cards. Once a rule is decided, it cannot be changed until the game is over.

8. The game is over when time is up or one player runs out of cards.

Materials: Set of Array Cards, paper and pencil

Players: 2

How to Play

1. Deal each player 10 Array Cards. Place 6 cards in the center of the table, dimensions side up. Place the deck of remaining cards nearby.

2. Place your cards in front of you, dimensions side up.

3. Take turns trying to complete a match—that is, to cover exactly an array in the center of the table with 2 or more smaller arrays. Look at your hand for a card that will cover part of an array in the center of the table. If no cards in your hand work, use a card from the center of the table to begin a match. If that doesn't work, pick the top card from the deck and try to play it. (If the deck card doesn't work, put it in your hand to use later.) At this point, it is the next player's turn.

4. If you use a card from the center of the table to cover another card, replace it with a card from your hand or from the deck. There should always be 6 cards in the middle of the table.

5. When you complete a match, collect it. The next player starts another match or completes a match started earlier in the same way. During the game, the players together keep a list of the dimensions of each larger array and the smaller ones that match it:

$$7 \times 6 = 3 \times 6 + 4 \times 6$$
$$42 = 18 + 24$$

6. If you run out of cards, take 4 cards from the deck. The game is over when all the cards have been matched or when no more matches can be made.

Another Set of Related Problems

Solve each of the following problems. Write about how you found the solution to 7×16.

7×8

7×6

7×10

7×16

This is how I solved 7×16:

_____

_____

_____

_____

_____

_____

_____

_____

_____

_____

Multiplication Clusters (page 1 of 5)

Sets A and B

Work with a partner to solve the problems in a cluster. After you work on several clusters, choose one to write about. To do that, tell how you used the answers of the first few problems in the cluster to help you find the answer to the last problem.

SET A

4×10

4×15

4×3

4×30

SET B

5×3

2×50

10×3

50×3

Multiplication Clusters (page 2 of 5)

Sets C and D

Work with a partner to solve the problems in a cluster. After you work on several clusters, choose one to write about. To do that, tell how you used the answers of the first few problems in the cluster to help you find the answer to the last problem.

SET C

2×5

3×5

10×5

30×5

32×5

SET D

5×7

10×7

4×25

20×7

25×7

Multiplication Clusters (page 3 of 5)

Sets E and F

Work with a partner to solve the problems in a cluster. After you work on several clusters, choose one to write about. To do that, tell how you used the answers of the first few problems in the cluster to help you find the answer to the last problem.

SET E

3×5

10×5

20×5

23 × 5

SET F

6×3

3×10

20×3

30×3

60 × 3

Multiplication Clusters (page 4 of 5)

Sets G and H

Work with a partner to solve the problems in a cluster. After you work on several clusters, choose one to write about. To do that, tell how you used the answers of the first few problems in the cluster to help you find the answer to the last problem.

SET G

2×5

2×30

3×30

2×100

2×95

SET H

2×7

2×10

2×30

2×70

Multiplication Clusters (page 5 of 5)

Sets I and J

Work with a partner to solve the problems in a cluster. After you work on several clusters, choose one to write about. To do that, tell how you used the answers of the first few problems in the cluster to help you find the answer to the last problem.

SET I

2×50

3×50

6×5

6×50

SET J

8×2

10×2

40×2

80×2

Problems About Our Class

Solve each word problem. Write how you solved the problem, or draw a picture. Use division notation to write an equation when it helps.

1. Suppose we want to arrange our desks into smaller groups of students. How many groups would we have if there were 5 desks in a group?

   How many groups would we have if there were 4 desks in a group?

2. There are 65 students going on a field trip. We need adult chaperones and vans. For every 9 students, we need one adult. Each van holds 10 people. How many adults do we need? How many vans do we need?

3. How many adults would we need if 4 classes like ours went on a field trip?

Recycling Problems

Solve each word problem. Write how you solved the problem, or draw a picture. Use division notation to write an equation when it helps.

1. Some fourth graders collected cans to recycle. They had a total of 46 cans. They needed to store them in boxes that hold 12 cans. How many boxes did they need?

2. On the first day of class-wide collections, students collected 90 cans. How many boxes did they need to store the cans? Each box held 12 cans.

3. The next day they collected 45 more cans. How many cans did they collect during the first two days? How many boxes did they use for storage?

Multiplication Clusters at Home

After you solve the problems, write about how you solved the last problem in the cluster.

Tell how you used the answers of the first few problems to help you find the answer to the last problem.

10×6

3×6

6×6

13×6

4×6

4×10

4×12

4×30

4×36

Two Ways to Solve a Problem

Solve this problem in two different ways:

$$27 \times 4$$

After each way, write about how you did it. Be sure to include:

- ■ what materials, if any, you used to solve this problem
- ■ how you solved it
- ■ an explanation of your thinking as you solved it

First Way:

Second Way:

Problems That Look Hard But Aren't

Choose a problem from the class list of Problems That Look Hard But Aren't. Solve the problem. Remember to show your work.

The problem I am solving is _____

Write why this problem looks hard but is not.

_____

_____

_____

_____

_____

_____

2	2	2	3
3	4	4	5
<u>6</u>	7	8	<u>9</u>
12	15	16	20
Wild Card	Wild Card	Wild Card	Wild Card

114

Ten-Minute Math
Arrays and Shares

Practice Pages

This optional section provides homework ideas for teachers who want or need to give more homework than is assigned to accompany the activities in this unit. The problems included here provide additional practice in learning about number relationships and in solving computation and number problems. For number units, you may want to use some of these if your students need more work in these areas or if you want to assign daily homework. For other units, you can use these problems so that students can continue to work on developing number and computation sense while they are focusing on other mathematical content in class. We recommend that you introduce activities in class before assigning related problems for homework.

101 to 200 Bingo　　This game is introduced in the unit *Mathematical Thinking at Grade 4*. If your students are familiar with the game, you can simply send home the directions, game board, Tens Cards, and Numeral Cards so that students can play at home. If your students have not played the game before, introduce it in class and have students play once or twice before sending it home. You might have students do this activity two times for homework in this unit.

Ways to Count Money　　This type of problem is introduced in the unit *Mathematical Thinking at Grade 4*. Here, three problem sheets are provided. You can also make up other problems in this format, using numbers that are appropriate for your students. Students find two ways to solve each problem. They record their solution strategies.

Solving Problems in Two Ways　　Solving problems in two ways is emphasized throughout the *Investigations* fourth grade curriculum. Here, we provide two sheets of problems that students solve in two different ways. Problems may be addition, subtraction, multiplication, or division. Students record each way they solved the problem. We recommend you give students an opportunity to share a variety of strategies for solving problems before you assign this homework.

How to Play 101 to 200 Bingo

Materials
- 101 to 200 Bingo Board
- One deck of Numeral Cards
- One deck of Tens Cards
- Colored pencils, crayons, or markers

Players: 2

How to Play

1. Each player takes a 1 from the Numeral Card deck and keeps this card throughout the game.

2. Shuffle the two decks of cards. Place each deck face down on the table.

3. Players use just one Bingo Board. You will take turns and work together to get a Bingo.

4. To determine a play, draw two Numeral Cards and one Tens Card. Arrange the 1 and the two other numerals to make a number between 100 and 199. Then add or subtract the number on your Tens Card. Circle the resulting number on the 101 to 200 Bingo Board.

5. Wild Cards in the Numeral Card deck can be used as any numeral from 0 through 9. Wild Cards in the Tens Card deck can be used as + or − any multiple of 10 from 10 through 70.

6. Some combinations cannot land on the 101 to 200 Bingo Board at all. Make up your own rules about what to do when this happens. (For example, a player could take another turn, or the Tens Card could be *either* added or subtracted in this instance.)

7. The goal is for the players together to circle five adjacent numbers in a row, in a column, or on a diagonal. Five circled numbers is a Bingo.

101	102	103	104	105	106	107	108	109	110
111	112	113	114	115	116	117	118	119	120
121	122	123	124	125	126	127	128	129	130
131	132	133	134	135	136	137	138	139	140
141	142	143	144	145	146	147	148	149	150
151	152	153	154	155	156	157	158	159	160
161	162	163	164	165	166	167	168	169	170
171	172	713	174	175	716	177	178	179	180
181	182	183	184	185	186	187	188	189	190
191	192	193	194	195	196	197	198	199	200

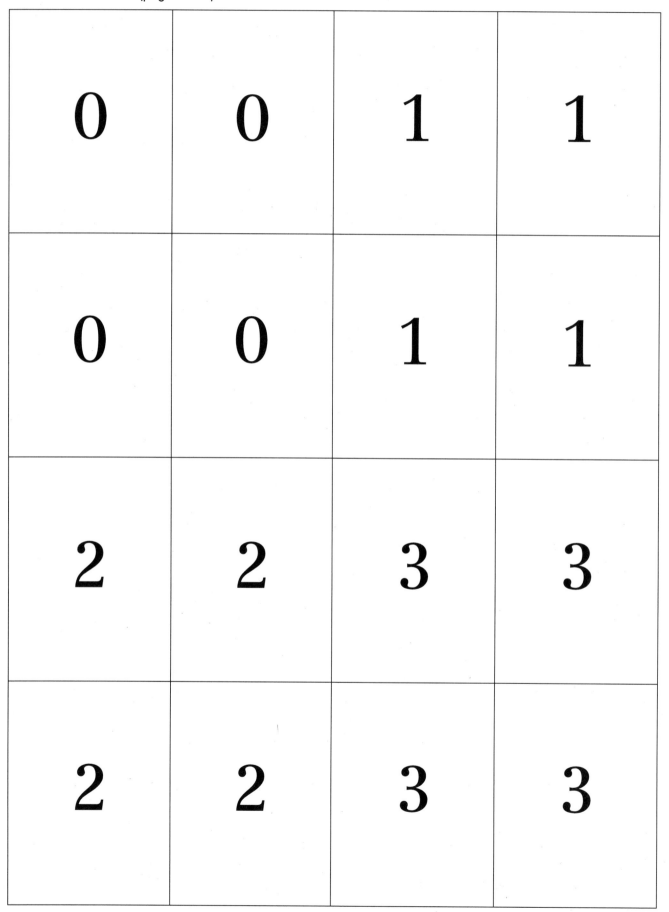

0	0	1	1
0	0	1	1
2	2	3	3
2	2	3	3

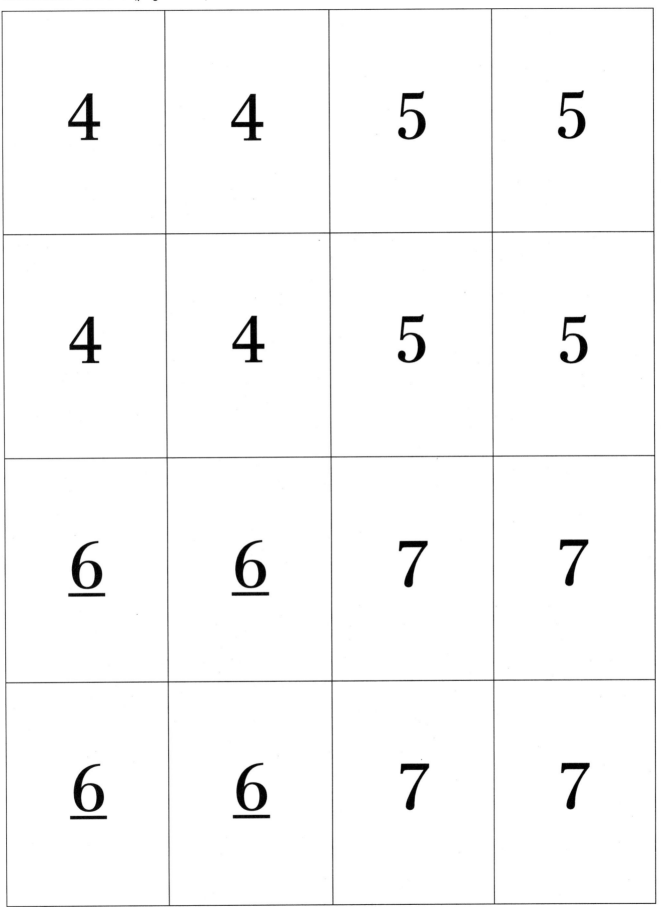

4	4	5	5
4	4	5	5
<u>6</u>	<u>6</u>	7	7
<u>6</u>	<u>6</u>	7	7

Practice Page
Arrays and Shares

8	8	9	9
8	8	9	9
WILD CARD	WILD CARD		
WILD CARD	WILD CARD		

Practice Page
Arrays and Shares

+10	**+10**	**+10**	**+10**
+20	**+20**	**+20**	**+20**
+30	**+30**	**+30**	**+40**
+40	**+50**	**+50**	**+60**
+70	**WILD CARD**	**WILD CARD**	**WILD CARD**

121

-10	**-10**	**-10**	**-10**
-20	**-20**	**-20**	**-20**
-30	**-30**	**-30**	**-40**
-40	**-50**	**-50**	**-60**
-70	**WILD CARD**	**WILD CARD**	**WILD CARD**

Practice Page
Arrays and Shares

Practice Page A

Find the total amount of money in two different ways.

    3 quarters
    6 pennies
    2 dimes
    10 nickels

Here is the first way I found the total amount of money:

Here is the second way I found the total amount of money:

Practice Page B

Find the total amount of money in two different ways.

    1 quarter
    6 nickels
    4 pennies
    3 dimes

Here is the first way I found the total amount of money:

Here is the second way I found the total amount of money:

Practice Page C

Find the total amount of money in two different ways.

      2 quarters
      5 nickels
      2 pennies
      1 dime

Here is the first way I found the total amount of money:

Here is the second way I found the total amount of money:

Practice Page D

Solve this problem in two different ways, and write about how you solved it:

84 + 26 =

Here is the first way I solved it:

Here is the second way I solved it:

Practice Page E

Solve this problem in two different ways, and write about how you solved it:

$$72 \div 9 =$$

Here is the first way I solved it:

Here is the second way I solved it:

127